VACCINATED

## ALSO BY PAUL A. OFFIT, M.D.

*Overkill*
*Bad Advice*
*Pandora's Lab*
*Bad Faith*
*Do You Believe in Magic?*
*Deadly Choices*
*Autism's False Prophets*
*The Cutter Incident*

# VACCINATED

*One Man's Quest to Defeat the World's Deadliest Diseases*

Paul A. Offit, M.D.

HARPER PERENNIAL

NEW YORK • LONDON • TORONTO • SYDNEY • NEW DELHI • AUCKLAND

HARPER PERENNIAL

A hardcover edition of this book was published in 2007 by HarperCollins Publishers.



First Smithsonian Books paperback edition published 2008.
First Harper Perennial edition published 2022.

*Designed by Renato Stanisic*

The Library of Congress Cataloging-in-Publication Data is available upon request.

ISBN 978-0-06-315761-3 (pbk.)

22 23 24 25 26 LSC 10 9 8 7 6 5 4 3 2 1

For Bonnie,

who made dreams come true,

and for our children, Will and Emily,

the two meteors streaking through our lives

**"And now," cried Max, "let the wild rumpus start."**

Maurice Sendak, *Where the Wild Things Are*

# CONTENTS

# FOREWORD

In November 2019, a bat coronavirus made its debut in the human population. By July 2021, more than 200 million people had been infected and four million killed. The virus was called SARS-CoV-2 and the disease COVID-19. One hundred years had passed since the world had experienced a pandemic of this magnitude.

There was, however, a way out: vaccines. For more than two hundred years, vaccines have had a remarkable record of success:

- Smallpox, a disease that killed more than 500 million people and changed the arc of European history, has been eliminated from the face of the earth.

- Rabies, a disease with a mortality rate of 100 percent, can now be prevented.

- Yellow fever, a virus that infected 50,000 people and killed 5,000 in Philadelphia in a single year, has been eliminated from the United States and Europe.

- Polio, a disease that caused 30,000 children to be paralyzed and 1,500 to die every year, was, by the late 1970s, eliminated from the United States. By 2020, a campaign by the

World Health Organization (WHO) had eliminated two of the three types of polioviruses that cause disease.

- Measles, a disease that caused 50,000 hospitalizations and 500 deaths a year, was, by 2020, no longer spreading in the United States. Worldwide, measles deaths have declined from 2.6 million to 200,000 a year.

- The mumps vaccine has virtually eliminated the most common cause of acquired deafness, causing many homes for the deaf to close their doors.

- Rubella (otherwise known as German measles), a virus that when it infected pregnant women caused as many as 20,000 cases of permanent birth defects and 5,000 spontaneous abortions every year, was, in 2005, eliminated from the United States. Worldwide, during the past twenty years, the rubella vaccine has reduced the number of cases from 670,000 to 49,000.

- The chicken pox vaccine has caused a 99 percent reduction in the 10,000 hospitalizations and one hundred deaths caused by the virus every year in the United States.

- A vaccine against *Haemophilus influenzae* type b (Hib), a bacterium that caused 25,000 cases of bloodstream infections and meningitis every year, has reduced that number to fewer than fifty. Vaccines against other bacteria, such as pneumococcus and meningococcus, which also cause bloodstream infections and meningitis, have dramatically reduced the incidence of those diseases.

- Before the hepatitis B vaccine, the virus infected 18,000 children younger than ten years of age every year in the United States, most destined to develop long-term liver disease (cirrhosis) and liver cancer. After the hepatitis B vaccine was routinely recommended for newborns in the early 1990s, the infection has been virtually eliminated in children. Further, thanks to the Bill and Melinda Gates Foundation, the hepatitis B vaccine has now been given to more than 85 percent of the world's population.

- A vaccine against human papillomavirus has resulted in an appreciable decline in the incidence of cervical cancer in several population studies.

- A vaccine against rotavirus, an intestinal virus that killed 500,000 babies every year in the world, is now saving hundreds of lives every day.

In short, because of vaccines, we live thirty years longer than we did a hundred years ago. Unfortunately, as is true in a war against any enemy, casualties occur on both sides. We learn as we go. And that knowledge often comes with a human price. Sadly, the story of vaccines is also littered with tragedy. For example:

In March 1942, the Office of the Surgeon General noted a growing incidence of jaundice among U.S. army personnel who had recently received a yellow fever vaccine that contained human serum as a stabilizing agent. The serum had been obtained from health care workers at Johns Hopkins Hospital in Baltimore, several of whom had a history of jaundice and one of whom was actively infected at the time of the donation. When the dust settled, 330,000 service members had been infected and one thousand had died from what would later be called hepatitis B virus. It was one of the worst single-source outbreaks of a fatal infection ever recorded.

In 1955, five pharmaceutical companies stepped forward to make Jonas Salk's polio vaccine. One, Cutter Laboratories of Berkeley, California, made it badly, failing to fully inactivate the virus. As a result, 120,000 children were inoculated with live, fully virulent polio virus; 40,000 were temporarily paralyzed, 164 were permanently paralyzed, and 10 were killed. It was arguably the worst biological disaster in American history.

In the early 1960s, a vaccine to prevent respiratory syncytial virus, a common cause of pneumonia in young children, was made by taking the virus and inactivating it with a chemical in much the same way that Jonas Salk made his polio vaccine. Researchers were hopeful that they now had a way to prevent a virus that killed five thousand babies every year in the United States. It didn't work out that way. Early studies found that children who were vaccinated were *more* likely to be hospitalized and *more* likely to die from pneumonia when later exposed to the virus than those who were never vaccinated. A similar problem occurred with two early versions of the measles vaccine, both of which were quickly removed from the market.

Knowing that the road to successful vaccines is often bumpy and occasionally fraught with dangers, the development of vaccines against SARS-CoV-2 has, for several reasons, been one of the most remarkable scientific achievements in the past two hundred years.

SARS-CoV-2 is an elusive virus. When public health officials in China billed SARS-CoV-2 as a respiratory virus that, like influenza, could cause severe and occasionally fatal pneumonia, they had severely underestimated its heinous nature. COVID-19 is far worse than anyone could have imagined. Within a year, the virus was discovered to have several clinical and pathological features that both surprised and confounded doctors and researchers. Specifically:

In addition to infecting the lungs, SARS-CoV-2 virus caused people to make an immune response against the lining of their blood vessels, thus causing inflammation (vasculitis). Because every organ in the body has a blood supply, every organ could be affected.

SARS-CoV-2 could cause people to lose their sense of taste and smell, often for weeks at a time.

Neuropathologists have detected SARS-CoV-2 virus in the brains of some people with COVID-19.

Children with mild or asymptomatic infections, who would rid themselves of the virus within a couple of weeks, would later be hospitalized with high fever and lung, liver, heart, and kidney disease. This postinfection phenomenon, called multisystem inflammatory syndrome in children (MIS-C), occurred in about one of every one thousand children infected and was occasionally fatal. Many children continued to experience symptoms more than two months later, so-called long-haulers. This same syndrome can also occur in adults.

SARS-CoV-2 virus causes inflammation of the heart muscle (myocarditis). A study performed in athletes in the Big Ten Conference found that one of every forty-three young men and women with COVID-19 had evidence of myocarditis.

No other respiratory virus does these things.

Perhaps worst of all, SARS-CoV-2 was constantly mutating, constantly trying to adapt itself to growth in people. The virus that first appeared in Wuhan, China, (called 2019-nCoV) was not the virus that left that country. That virus, called D614G, was the first significant mutation (or variant). And it was far more contagious than the original virus. Indeed, it was the D614G variant that swept across Asia, Europe, and the United States, killing millions of people, only to be replaced by yet another variant, the alpha variant, which was even more contagious. But SARS-CoV-2 wasn't finished mutating. The alpha variant was later replaced by the even more contagious delta variant. As the variants became more and more contagious, a greater and greater percentage of the population needed to be vaccinated to stop the spread of the virus.

The first vaccine strategies available to quell this virus had never been used before. Instead of giving people a live, weakened form of SARS-CoV-2 virus (similar to the measles, mumps, rubella, chicken pox, and rotavirus vaccines), or a killed form of the virus (similar to the rabies or inactivated polio vaccines), or a purified single protein from the virus (similar to the hepatitis B and human papillomavirus vaccines), people were inoculated with the gene that was coded for the surface (or spike) protein of SARS-CoV-2. These novel, genetic vaccines contained either a naked piece of messenger RNA (mRNA) or a Trojan horse virus (vectored virus) that delivered the SARS-CoV-2 gene into the cell. People vaccinated with these genetic vaccines would make the SARS-CoV-2 spike protein in their own cells and then make antibodies to the spike protein. The birth of the genetic era of vaccination.

In addition to the novelty of these vaccines, several other aspects of what would soon be a worldwide, mass vaccination program worried people:

1. The speed with which the vaccine was made. SARS-CoV-2 virus was isolated and characterized in January 2020. Less than one year later, the first two mRNA vaccines had been tested in large trials. These COVID-19 vaccines were the fastest vaccines ever made. (The previous record from isolation of a virus to a commercially available product was the mumps vaccine, which took four years.)

2. The mechanism by which these vaccines were approved by the Food and Drug Administration (FDA). Normally, pharmaceutical companies submit their products for licensure, which takes on average about ten months. This time, because of a rapidly spreading, deadly pandemic, the FDA chose a different mechanism, called Emergency Use Authorization (EUA): a lower bar. The time from submission to approval was weeks, not months.

3. The language surrounding EUA approval. Phrases like "Operation Warp Speed," "the race for a vaccine," and "who was going to be the first to cross the finish line" frightened people who thought vaccine timelines were being truncated or, worse, that safety guidelines were being ignored.

Taken together, in less than a year, a new virus had caused several unanticipated clinical and pathological problems that were met with vaccine strategies that had never been used before, that were developed more quickly than any vaccine ever made, and that had been approved by a mechanism (EUA) that was clearly less stringent than the typical licensure process. Everyone held their breath, assuming that it would only be a matter of time before another vaccine tragedy occurred.

It never happened. The mRNA vaccines made by Pfizer and Moderna and the Trojan horse viral vaccines made by Johnson & Johnson were remarkably effective, preventing more than 90 percent of severe infections in all age groups and in all those who were at highest risk for the disease. The result was far better than anyone could have predicted or imagined. Also, the vaccines were safe. But they weren't without issue; the Johnson & Johnson vaccine was a very rare cause of clotting, and the mRNA vaccines were a very rare cause of myocarditis. But their huge benefits outweighed their small risks.

Once these vaccines were in hand, stopping the pandemic centered on two challenges. First, there was the challenge of getting the vaccine into countries that could least afford them. In late 2021, of the 195 countries in the world, most hadn't given a single dose of the COVID-19 vaccines. Second, and most depressing, several developed-world countries, including the United States, were finding that a substantial percentage of the population, as much as 30 percent, were simply refusing to be immunized. As a result, the virus was continuing to spread and continuing to generate more contagious variants, making it harder and harder to contain.

None of this, however, should have been surprising. Due to a strident antivaccine movement, people in many developed-world countries had for several decades been choosing not to vaccinate themselves or their children. The one person who was hit hardest by this rejection was the man who had created nine of the fourteen vaccines currently given to infants and young children. A man who was the first to predict an influenza pandemic and to make a vaccine in advance of its entry into the United States. A man who won every major award given to medical researchers in the United States, including the National Medal of Science presented by President Ronald Reagan. A man whose work is estimated to save about eight million lives a year. And a man who most people have never heard of—Maurice Hilleman. In many ways, as you'll see in the pages that follow, the story of vaccines is his story.

Paul A. Offit, MD
July 2021

# PROLOGUE

Scientists aren't famous. They never endorse products or sign autographs or fight through crowds of screaming admirers. But at least you know a few of their names, like Jonas Salk, the developer of the polio vaccine; or Albert Schweitzer, the missionary who built hospitals in Africa; or Louis Pasteur, the inventor of pasteurization; or Marie Curie, the discoverer of radiation; or Albert Einstein, the physicist who defined the relationship between mass and energy. But I'd bet not one of you knows the name of the scientist who saved more lives than all other scientists combined—a man who survived Depression-era poverty; the harsh, unforgiving plains of southeastern Montana; abandonment by his father; the early death of his mother; and, at the end of his life, the sad realization that few people knew who he was or what he had done: Maurice Hilleman, the father of modern vaccines.

Hilleman's science followed a long, rich tradition.

In the late 1700s Edward Jenner, a physician working in southern England, made the world's first vaccine. Jenner found that he could protect people from smallpox—a disease that has claimed five hundred million victims—by injecting them with cowpox, a related virus.

One hundred years passed.

In the late 1800s Louis Pasteur, a chemist working in Paris, made the world's second vaccine. Pasteur's vaccine, made by drying spinal cords from infected rabbits, prevented the single most deadly infection of man—rabies. Only a handful of people have ever survived rabies without receiving a rabies vaccine.

During the first half of the twentieth century, scientists made six more vaccines. In the 1920s French researchers found that bacteria made toxins and that toxins treated with chemicals could be used as vaccines. These observations led to vaccines against diphtheria, tetanus, and, in part, whooping cough. In the 1930s a researcher at the Rockefeller Institute in New York City made a yellow fever vaccine by growing the virus in mice and chickens. In the 1940s Thomas Francis, working at the University of Michigan, made an influenza vaccine by growing the virus in eggs and killing it with formaldehyde. And in the 1950s Jonas Salk and Albert Sabin, using monkey kidneys, made polio vaccines that eventually eradicated polio from the Western Hemisphere and much of the world.

The second half of the twentieth century witnessed an explosion in vaccine research and development, with vaccines to prevent measles, mumps, rubella (German measles), chickenpox, hepatitis A, hepatitis B, pneumococcus, meningococcus, and *Haemophilus influenzae* type b (Hib). Before these vaccines were made, Americans could expect that every year measles would cause severe, fatal pneumonia; rubella would attack unborn babies, causing them to go blind or deaf or become developmentally disabled; and Hib would infect the brain and spinal cord, killing or disabling thousands of young children. These nine vaccines virtually eliminated all of this suffering and disability and death. And Maurice Hilleman made every one of them.

In October 2004, doctors told Hilleman that he had an aggressive form of cancer, one that had already spread to his lungs and would likely soon overwhelm him. During the six months before he died, Hilleman talked to me about his life and work. This book—the story of the triumphs, tragedies, controversies, and uncertain future of modern vaccines—is largely his story.

# THE TIME CAPSULE

The dedication of the National Millennium Time Capsule marked the end of the twentieth century. To fill the capsule, First Lady Hillary Clinton sent an invitation to four hundred presidential and congressional medal winners. "You have been recognized for your contributions to the nation," she wrote. "Now I would like to ask you for another contribution. If you could choose just one item or idea to represent America at the end of the twentieth century, and to be preserved for the future, what would it be?"

Ray Charles submitted a pair of his sunglasses.

Wilma Mankiller, chief of the Cherokee Nation, submitted the eighty-five-letter Cherokee alphabet, hoping that her language would still be spoken a hundred years from now.

Hans Liepmann, a mathematician and scientist, submitted the first transistor—developed by Bell Telephone Laboratories—to mark the beginning of the electronic age.

Historian David McCullough submitted a borrower's card from the Boston Public Library, the first public library to let readers take books home.

President Ronald Reagan submitted a piece of the Berlin Wall to represent one country's choice of democracy over communism.

Filmmaker Ken Burns submitted an original recording of Louis Armstrong's "West End Blues."

Ernest Green, an African-American student caught on September 2, 1957, in a confrontation between Arkansas governor Orval Faubus and armed national guardsmen about his admission to an all-white public school, submitted his diploma from Little Rock Central High School.

Others submitted a microchip, the first artificial heart, a piece of the Transoceanic Cable, a copy of John Steinbeck's *The Grapes of Wrath*, a film of *Apollo II* landing on the moon, broadcasts from the Metropolitan Opera, a piece of Corning Ware, a film of Jackson Pollock creating one of his drip paintings, a copy of the genetic code, Bessie Smith's recording of "Nobody in Town Can Bake a Sweet Jellyroll Like Mine," and photographs of the earth from space, the atomic bomb's mushroom cloud over Hiroshima, and American servicemen liberating prisoners from a Nazi concentration camp in Buchenwald.

The ceremonial placement of items in the time capsule took place on Friday, December 31, 1999, a brisk, windy, winter day in Washington, D.C. President Bill Clinton and First Lady Hillary Clinton spoke at the event, and ten thousand people lined the streets near the National Mall to watch. "It was, after all, the transistor that launched the Information Age and enabled man to walk on the moon," said Mrs. Clinton. "It was Satchmo's trumpet that heralded the rise of jazz and of American music all over the world. And it was a broken block of concrete covered in graffiti from the Berlin Wall that announced the triumph of democracy over dictatorship." Bill Clinton expressed his hopes for the future. "There is not a better moment to reflect on our hopes and dreams, and the gifts we want to leave to our children," he said.

Another man was on the platform that day: Maurice Hilleman. Few in attendance recognized him. Eighty years old, bent slightly forward, Hilleman slowly, cautiously padded over to the microphone, said a few words, and reached down to place his artifact into the capsule: a block of clear plastic six inches long, two inches high, and two inches deep. Embedded in the plastic were several small vials.

Although it was never mentioned during the ceremony, Americans now lived thirty years longer than they had when the century began. Some of this increase in longevity was caused by advances such as antibiotics, purified drinking water, improved sanitation, safer workplaces, better nutrition, safer foods, the use of seat belts, and a decline in smoking. But no single medical advance had had a greater impact than what was contained in Hilleman's vials—vaccines.

Four years after the time capsule ceremony, a reporter asked Hilleman how it felt to stand next to the president of the United States on the century's final day, how it felt to participate in a moment that crystallized his career. A taciturn, gruff, humble man, Hilleman was uncomfortable taking credit for what he had done, uncomfortable looking back. "Chilly," he said.

# VACCINATED

# CHAPTER 1

## "My God: This Is the Pandemic. It's Here!"

*"I had a little bird, and its name was Enza,*
*I opened the window, and in-flew-Enza."*

CHILDREN'S RHYME DURING THE 1918 FLU PANDEMIC

In May 1997 a three-year-old boy in Hong Kong died of influenza. His death wasn't unusual. Every year in every country in every corner of the world healthy children die of the disease. But this infection was different. Health officials couldn't figure out what type of influenza virus had killed the boy, so they sent a sample of it to the Centers for Disease Control and Prevention (CDC) in Atlanta. There, researchers found that this particular virus had never infected people before. A few months passed. The rare influenza virus infected no one else—not the boy's parents, or his relatives, or his friends, or his classmates. Later, the CDC sent a team of scientists to Hong Kong to investigate. Crowded into a wet market, where local farmers slaughtered and sold their chickens, they found what they were looking for—the source of the deadly virus. "The people here like their chickens fresh," one investigator said. "Hygiene consists of a douse with cold water. [One day] we saw a bird standing up there, pecking away at its food, and then very gently lean over, slowly fall over, to lie on its side, looking dead. Blood was trickling from [its beak]. It was a very unreal, bizarre situation. I had never seen anything like it." The disease spread to another chicken and another.

The strain of influenza virus that infected birds in Southeast Asia was particularly deadly, killing seven of every ten chickens. On December 30, 1997, Hong Kong health officials, in an effort to control the outbreak of bird flu before it spread to more people, slaughtered more than a million chickens. Still the virus spread. Bird flu attacked chickens in Japan, Vietnam, Laos, Thailand, Cambodia, China, Malaysia, and Indonesia. Then, to the horror of local physicians, the virus infected eighteen more people, killing six—a death rate of 33 percent. (Typically influenza kills fewer than 2 percent of its victims.) Soon the virus disappeared. Officials waited for an outbreak the following year, but none came. And it didn't come the year after that or the year after that. The virus lay silent, waiting.

In late 2003, six years after the initial outbreak, bird flu reappeared in Southeast Asia. This time health officials found it even harder to control. Again, the virus first infected chickens. Officials responded by slaughtering hundreds of millions of them. Despite their efforts, bird flu.spread from chickens to ducks, geese, turkeys, and quail. Then the virus spread to mammals: first to mice, then to cats, then to a tiger in a Thai zoo, then to pigs, then to humans. By April 2005, bird flu had infected ninety-seven people and killed fifty-three—a death rate of 55 percent.

By September 2006 the virus had spread from birds in Asia to those in Europe, the Near East, and Africa. Two hundred fifty people living close to these birds got sick, and 146 of them died. International health officials feared that the appearance of bird flu in Southeast Asia signaled the start of a worldwide epidemic (pandemic). One later remarked, "The clock is ticking. We just don't know what time it is."

Health officials feared an influenza pandemic because they knew just how devastating pandemics could be. During the pandemic of 1918 and 1919—the one called the last great plague—influenza infected five hundred million people, half the world's population. The virus, which traveled to virtually every country and territory in the world, hit the United States particularly hard. In a single month, October 1918, four hundred thousand Americans died of influenza. Influenza typically kills the most vulnerable members of the population, the sick and the elderly. But the 1918 virus was different: it killed healthy young adults. In one year the average life span of Americans in their twenties and thirties decreased by 25 percent. When it was over, the 1918 pandemic—the most devastating outbreak of an infectious disease in medical history—had killed between fifty million and one hundred million people worldwide, all within a single year. In comparison, since the 1970s the AIDS pandemic has killed thirty-five million people.

Pandemics of influenza are inevitable. During the past three hundred years, the world has suffered ten of them, about three per century. No century has ever avoided one. But despite their frequency and reproducibility, only one man has ever successfully predicted an influenza pandemic and done something about it.

HIS NAME WAS MAURICE HILLEMAN. BORN SATURDAY MORNING, August 30, 1919, during the worst influenza pandemic in history, Hilleman was the eighth child of Anna and Gustave Hillemann. (Because of intense anti-German sentiment following the First World War, Hilleman's parents deleted the second *n* on his birth certificate.) Devoutly religious, Anna and Gustave named him and his sister (Elsie) and all of his brothers (Walter, Howard, Victor, Harold, Richard, and Norman) after heroic characters in the Elsie Dinsmore books, stories of Christian faith popular in the late 1800s. The birth took place in the family's home on the banks of the Tongue and Yellowstone Rivers, near Miles City, Montana.

After Maurice's birth, and to the surprise of the homeopath who delivered him, a second child, Maureen, was also born, still and lifeless. The doctor tried desperately but unsuccessfully to revive her. He cupped his hands around her back and, using only his thumbs, periodically pushed down on her tiny chest. At the same time, he tried to breathe air into her lungs. It was no use. From the corner of the room, Anna Hillemann quietly watched the doctor try to save

*Hilleman's birthplace, Custer County, Montana, circa 1919.*

her baby daughter. When she learned that Maureen was dead, Anna closed her eyes, saying nothing. Gustave buried Maureen the following day, August 31.

Hours after the delivery, while she was holding her infant son, Anna's body stiffened, her eyes rolled up in her head, foam collected at the corners of her mouth, and her arms and legs twitched rhythmically and unstoppably. The seizure was the first of many; for several hours after each one she lay unconscious in her bed. The doctor declared that Anna was suffering from eclampsia, a disease unique to pregnant women, caused by a progressive, unrelenting swelling of the brain. Anna knew that she was dying. So she called her husband, Gustave; his brother, Robert; and Robert's wife, Edith, to her bedside. She asked that the older boys remain on the family farm with Gustave; that Elsie, Richard, and Norman live with her relatives in Missouri; and that her new baby, Maurice, be raised by Robert and Edith, who lived just down the road. Anna felt bad for the childless couple, so she gave them her infant son. Two days after the birth, Anna Hillemann, like her baby daughter, died. But before she died, Anna made one more request. Two days later Gustave obliged, exhuming baby Maureen and burying her in her mother's arms. Maurice was the only one who survived the birth. "I always felt that I cheated death," he said.

Although he lived with Robert and Edith in a house separate from his brothers and sister—all later reunited with Gustave—Maurice worked on the family farm, the Riverview Garden and Nursery. "At one time [our farm] provided an escape for thieves and outlaws who were [being] pursued by vigilante posses from Miles City," recalled Hilleman. "There was still a tall cottonwood on the high ground that carried a hangman's noose in its branches." Hilleman remembered life on the farm: "We sold anything that people would buy: potatoes, tomatoes, cabbage, lettuce, radishes, corn, squash, cottage cheese, dressed chickens, hatching eggs, eating eggs, and pumpkins. We made brooms from the straw of sorghum sugar—brooms that lasted forever. In Miles City, we did landscaping, tree surgery, and sprayed the trees for worms and insect manifestations. We even did landscaping for the whorehouses. [Prostitution was legal in Miles City.] We raised perennials and annuals and sold flowers, usually on Sundays, to the local florist. When I was old enough to tell the difference between a weed and a plant, I was sent out into the sun, working from sunup to sunset. My jobs were to pick berries, bring in the horses, pump water, feed and water the chickens, collect eggs, keep the chicken coop real clean, shovel shit from the roost, and pick beans. I worked only during the summer months because during school the frost killed everything."

*Maurice Hilleman, circa 1920.*

"Everyone had to earn their keep in Montana," recalled Hilleman's eldest daughter, Jeryl. "This was the eastern plains, where life was very, very brutal: broiling hot summers, freezing winters, snow

*Maurice Hilleman, circa 1923.*

drifts above [your] head. When he was four years old, [my father] was sent down to sell strawberries. He was told what [price] to ask for, but he didn't really sell anything. As the day went by, the strawberries got softer and less attractive, and he ended up selling them for a fraction of what he was supposed to. And he got seriously punished for it. Even at the age of four there was no mercy. He had a tough life growing up."

By the time he was only ten years old, Hilleman had survived near drowning, an unyielding freight train, and diphtheria. "In Montana you were responsible for yourself," he recalled. "Nobody was out looking after you. At flood time the Yellowstone River, which comes down from the mountains, takes out trees and houses. One day a hobo floated downstream in a small flatbed boat. He sold us this little piece of junk, this wooden thing, for one dollar. So my brother and I rowed down the Yellowstone. Now this was like rowing over a falls—like going over Niagara Falls—and there were these big cottonwood trees turning over, and junk and everything floating downstream. I couldn't swim. I barely made it back to the bank of the river, full of mud." Hilleman ran back to his aunt, breathless and muddy. He told her how he had almost drowned. Edith looked up, stared at Maurice for a moment, and went back to washing her family's clothes, saying nothing. "She was Lutheran," recalled Hilleman. "She figured that when your time had come, your time had come."

Another of Hilleman's misadventures involved the unexpected appearance of a freight train on a narrow bridge over the Tongue River. "The Milwaukee Railroad ran two trains through Miles City, the Olympian from Chicago to Seattle and the Columbian. Every morning we took our bicycles across a small bridge over the Tongue River. I always looked to make sure that the Olympian wasn't coming and that I saw the tail end. Well, we started on the bridge and we were about two-thirds across and, Jesus Christ, here comes the freight train. Those bastards had sent through an extra freight train. I was with my brother Norman. You know what it costs to stop a freight train, about a million bucks. It ruins the rails. [The conductor] is blowing his whistle, and he [doesn't] stop the brakes for anything. I looked back and the bridge was shaking. So we ran with our bicycles and we got to the end of the bridge and I threw my bicycle

*Maurice Hilleman, 1925.*

on the ground and we jumped. We had about one or two seconds before the train caught us."

When he was eight years old, Hilleman almost died of suffocation caused by diphtheria: "I was proclaimed near dead so many times as a kid. They always said I wouldn't last till morning."

Although legally adopted and raised by his aunt and uncle, Hilleman lived within a hundred yards of his natural father, a strict Lutheran. He railed against his father's conservative fundamentalism: "People in Montana were good people. It gets below zero. They helped each other out. They were a community. The church maintained law and order and provided a social structure and adhesiveness that was very important on the western front. It was a decent, ordered society. But I just couldn't buy all of the mythology. And I didn't want to be strapped down by the church's dogma." As a boy, Hilleman found solace in the pages of Charles Darwin's *The Origin of Species*, which he read and reread. "I was enthralled by Darwin because the church was so opposed to him," he recalled. "I figured that anybody who could be so universally hated had to have something good about him." Hilleman read everything he could find about science and the great men of science. His hero—and the hero of many Montana schoolchildren—was Howard Taylor Ricketts.

In the early 1900s, a mysterious disease attacked residents of

*Custer County High School, 1937*

*Maurice Hilleman, graduation day, Montana State University, 1941.*

the Bitterroot Valley in western Montana, causing high fever, intense headaches, muscle pain, low blood pressure, shock, and death. Montana's governor, Joseph Toole, called on Howard Ricketts to find the cause. A Midwesterner and a graduate of Northwestern University, Ricketts immediately recruited students from Montana State University in nearby Bozeman to help. Many of them later died of the disease. Much to his surprise, Ricketts found that the deadly infection was caused by a bacterium found in ticks. "He was a god out there," recalled Hilleman. Today the bacterium is called *Rickettsia rickettsii*, and the disease is called Rocky Mountain spotted fever.

As a teenager attending Custer County High School in Miles City, Hilleman landed a job as assistant manager at the J. C. Penney store, helping "cowpokes pick out chenille bathrobes for their girlfriends." In Depression-era Montana, this was a highly sought after position, and it ensured Hilleman's future. But one of Hilleman's brothers suggested that Maurice forget about J. C. Penney and go to college. "If you lived in Miles City and you were smart, you went to Concordia College and then to the seminary to be ordained as a Lutheran preacher. But I wasn't going to do that." So Hilleman applied for and won a full scholarship from Montana State University.

In 1941 he graduated first in his class, having majored in chemistry and microbiology.

After graduation, Hilleman wanted to go to medical school but again couldn't afford it. "I couldn't see any possibility of doing that," he said. "You had to put yourself through residency. Where was I going to get the money?" So Hilleman applied to ten graduate schools, hoping to eventually get a doctorate in microbiology. "I came from Montana State University, this small agricultural school. These people would see a letter from some cowboy in Montana, [and] I assumed that it would be in the wastebasket pretty quick." On the top of Hilleman's list was the University of Chicago. "To a westerner the United States ended in Chicago," recalled Hilleman. "Chicago was the mecca. The great university there was the intellectual center at the time." Hilleman was accepted by all ten schools, each offering a full scholarship: "I got into Chicago. I got into them all. Do you believe that? I was on top of the mountain."

Life in Chicago wasn't much easier than life on the farm. "He weighed 138 pounds," recalled Hilleman's wife, Lorraine. "He ate one meal a day; that was all he could afford. And I know that he slept with bedbugs, too. He would put bars of soap around to catch [them]." Chicago's academic style was also challenging. Hilleman recalled, "The Chicago system, as far as the professor was concerned, was 'Don't bother me and let me know when you discover something.'" Hilleman struggled to find a research project, eventually settling on chlamydia, a sexually transmitted pathogen that scientists thought was a virus. (Chlamydia infects about three million people in the United States every year, and, because it scars the fallopian tubes, causes infertility in tens of thousands of women.) Within a year, Hilleman found that chlamydia wasn't a virus at all; it was a small, unusual bacterium that, unlike other bacteria, grew only inside of cells. Hilleman's finding eventually led to a treatment for the disease. For his efforts, he received an award for "the student presenting the best results of research in pathology and bacteriology." Myra Tubbs Ricketts, the widow of Hilleman's childhood hero, had endowed the award. Hilleman remembered, "You see how things tie together in your life: Howard Taylor Ricketts!"

In 1944 Maurice Hilleman came to a crossroads. He had just finished his graduate studies in Chicago. Now he was expected to take his place among the academic elite as a teacher and researcher. But Hilleman wanted to work for the pharmaceutical company E. R. Squibb in New Brunswick, New Jersey. His mentors made it clear to him that working for a pharmaceutical company wasn't an option. "I left Chicago under significant duress," recalled Hilleman. "Because Chicago at that time was such an intellectual center of biology, no one went into industry. When you graduated from the University of Chicago, you would be announced [into the field of science]. I was not allowed to look for a job in industry." But Hilleman had tired of academia. "What am I supposed to do? I was told that you could teach or you could do research. I said I wanted to go into industry because I'd learned enough about academia. I wanted to learn something about industrial management. I came off a farm. We had to do marketing. We had to do sales. I wanted to do something. I wanted to make things!" Hilleman's decision irked his professors. So they added one more hurdle before allowing him to graduate: a French exam. "I spent six months learning French," he recalled. "Every day I learned ten pages of philosophic French and a hundred idioms and analogies. I passed the test." Reluctantly, Hilleman's mentors abandoned their protests. "Now you can go into industry," they said. At Squibb, Hilleman learned how to mass-produce influenza vaccine.

Four years later, in the late spring of 1948, he arrived at the Walter Reed Institute in Washington, D.C. His assignment at Walter Reed was to learn everything he could about influenza and to prevent the next pandemic. Confident, tall, and handsome, Hilleman commanded the respect of his research team with intellect, profanity, and humor.

Established on May 1, 1909, the Walter Reed Army Medical Research Institute supported research on any infection that could influence the outcome of wars. History supported its mission.

During Britain's occupation of India in the early 1800s, one third of its troops died of cholera. During the Crimean and Boer Wars, in the mid- and late 1800s, more British troops died of dysentery than in battle. During the First World War, typhus, a bacterial infection spread by lice, infected hundreds of thousands of Serbs and Russians. And during the Second World War, influenza killed thousands of American soldiers.

But for demonstrating how infections can change the outcome of wars, no war matched the Spanish conquest of Mexico in the sixteenth century. With an army of only four hundred men, Hernando Cortez conquered an Aztec civilization of four million. Cortez didn't defeat the Aztecs because his men were braver (the Aztecs were fierce, noble fighters) or because he was more adept at recruiting other Indian tribes to join him (allies joined Cortez only after they were sure he would win) or because he had more guns and horses (the guns were crude and the horses were of limited value). So what was it? What caused an Aztec civilization of millions to lay down its arms and wholly, unequivocally surrender to a few hundred Spanish invaders? The answer was an infection that had been circulating in Europe for centuries but had never crossed the Atlantic Ocean: smallpox. Within one year of the Spanish invasion, the smallpox plague had killed millions. The Aztecs interpreted the plague as punishment, convinced that their invaders enjoyed divine favor. "The religions, priesthoods, and way of life built around the old Indian gods could not survive such a demonstration of the superior power of the God the Spaniards worshipped," wrote William McNeill, author of *Plagues and Peoples*. "Little wonder, then, that the Indians accepted Christianity and submitted to Spanish control so meekly. God had shown Himself on their side, and each new outbreak of infectious disease imported from Europe, and soon from Africa as well, renewed the lesson."

In the late 1940s two institutions monitored strains of influenza virus that circulated in the world: the Walter Reed Institute and the newly formed World Health Organization (WHO) in Geneva, Switzerland. "I was in charge of the central laboratory for the military worldwide surveillance for early detection of [pandemic] viruses," recalled Hilleman. "And in 1957 we all [initially] missed it. The military missed it and the World Health Organization missed it."

On April 17, 1957, while sitting in his office, Hilleman read an article in the *New York Times* titled "Hong Kong Battling Influenza Epidemic." "I saw an article that said that there were twenty thousand people lined up being taken to the dispensaries. And children with glassy-eyed stares, tied to their mother's backs, were waiting to be seen." Public health officials estimated that the virus had infected two hundred fifty thousand people, 10 percent of Hong Kong's population. Hilleman put down the paper: "My God," he said, "This is the pandemic. It's here!"

The next day Hilleman cabled the army's 406th Medical General Laboratory in Zama, Japan. He asked the staff to find out what was going on in Hong Kong. A medical officer sent to investigate eventually found a navy serviceman who had been exposed to the virus in Hong Kong, gotten back on his ship, returned to Japan, and become ill. The officer asked the young man to gargle with salt water and spit into a cup, hoping to capture the virus.

The specimen reached Hilleman on May 17, 1957. For five days and nights he worked to determine whether the influenza virus circulating in Hong Kong could be a pandemic strain. Hilleman took an incubating hen's egg, cut a small window in the shell, and injected the egg with throat washings from the navy serviceman. Influenza virus grew readily in the membrane surrounding the chick embryo. He harvested the virus-containing fluid, purified it, and added sera from members of the American military, hundreds in all. No one had antibodies to the new virus. (Serum [plural "sera"] is the fraction of blood that contains antibodies. Antibodies are proteins made by the immune system to neutralize invading viruses and bacteria.) Hilleman then tested sera from hundreds of civilians in the United States—again no antibodies. He couldn't find one person whose immune system had ever confronted this particular strain of influenza virus before.

To confirm his findings, Hilleman sent the virus to the WHO, the

*Maurice Hilleman and the pandemic influenza team,*
*Walter Reed Army Medical Research Institute, 1957.*

U. S. Public Health Service, and the Commission on Influenza of the Armed Forces Epidemiological Board. Each of these organizations tested sera from adults throughout the world. Only a handful of people—in the Netherlands and the United States—had antibodies to the virus. All were elderly men and women in their seventies and eighties who had survived the influenza pandemic of 1889–1890 that had killed six million people. The virus that had caused the 1889 pandemic had disappeared quickly and mysteriously. Now it was back. And no one had antibodies to stop it.

Hilleman knew that he was studying a strain of influenza virus that could sweep across the world unchecked. He also knew that the only way to stop it would be to make a vaccine—and to make it quickly. On May 22, 1957, he sent out a press release claiming that the next influenza pandemic had arrived. "I had a very difficult time getting anybody to even believe this," remembered Hilleman. "I got a call from Macfarlane Burnet [an Australian immunologist], and he said that you cannot say that this virus is different. Joe Bell [from the U. S. Public Health Service] didn't believe it. He said, 'What pandemic? What influenza?' How [could] they live in this world and be so stupid?"

Hilleman extended his prediction. Not only did he believe that the virus circulating in Hong Kong—now called Asian flu—would spread throughout the world, but also he believed that it would enter the United States in the first week of September 1957. "And when I put out a press release stating that there was a pandemic going to come on the second or third day of the opening of school, I was declared crazy," he said. "But it came, on time."

While Hilleman was reading the article in the *New York Times* about the influenza outbreak in Hong Kong, the virus was already well on its way to starting a pandemic. The first case of Asian flu had appeared in February 1957 in Guizhou Province in southwestern China. By March it had spread to Hunan Province. Chinese refugees had then carried the virus to Hong Kong. By the end of April, Asian flu had spread to Taiwan, and by early May it had traveled to Malaysia and the Philippines. About two hundred children in the Philippines died of the infection. By late May the virus had spread to India, South Vietnam, and Japan. Mild epidemics occurred on ships of the U. S. Navy, which eventually transported the virus to the rest of the world. Hilleman recalled, "[Navy servicemen] would have an epidemic on board ship, and they would go into town in these ports of call and mix it up with the bar girls and really spread it throughout the Far East."

Hilleman had shown that Asian flu, which hadn't circulated in the world for seventy years, was back. Thomas Francis, head of the Influenza Commission for the American military, didn't believe him and refused to work on a vaccine. Hilleman remembered, "The military wasn't able to get the Influenza Commission to move. They sent the strains of virus down to Francis's lab, and he said, 'We'll look at it.' [One night] I knew that Tommy Francis was eating [dinner] in town at the Cosmos Club [in Washington, D.C.]. So I sat right beside the door till he walked in. 'Tommy Francis, I've got to show you something because you're making a huge mistake. We don't have time to fool around.' He looked through the data and said, 'My God, it's a pandemic virus.'"

Hilleman sent samples of Asian flu virus to six American-based companies that made influenza vaccine. He figured that if he were to have any hope of saving American lives, he would have to convince companies to make and distribute vaccine in four months. Influenza vaccine had never been made that quickly. Hilleman sped up the process by ignoring the Division of Biologics Standards, the principal vaccine regulatory agency in the United States. "I knew how the system worked," he said. "So I bypassed the Division of Biologics Standards, called the manufacturers myself, and moved the process quickly. The most significant thing I told them was to please advise their chicken producers not to kill the roosters or we won't have fertile eggs." Hilleman knew that the production of millions of doses of influenza vaccine would require hundreds of thousands of eggs a day and—thanks to his background in raising chickens—that farmers typically killed their roosters late in the hatching season.

As Hilleman had predicted, in September 1957 Asian flu entered the United States from both coasts. The first laboratory-proven cases occurred aboard naval vessels in San Diego, California, and Newport, Rhode Island. A San Diego girl started the first outbreak when she carried the virus to an international church conference in Grinnell, Iowa. The second outbreak occurred in Valley Forge, Pennsylvania. "A contingent of Boy Scouts [had come] from Hawaii," recalled Hilleman, "and they [were] put on the trains and moved from the West Coast all the way back to Valley Forge. An epidemic broke out, and [public health officials] tried to separate the cars and do all sorts of things, but those kids were just laid out everywhere. When they got them to Valley Forge, they put them all in pup tents, and the outbreak stopped."

Pharmaceutical companies made the first lots of Asian influenza vaccine in June 1957, and vaccinations began in July. By late fall, companies had distributed forty million doses. Asian flu quickly spread across the United States. The National Health Survey estimated that twelve million people were sick with influenza during the week beginning October 13. Within a few months influenza had infected twenty million Americans. Although the 1957 pandemic killed only a fraction of those killed during the 1918 pandemic, the two pandemics shared one sad feature: the disease disproportionately killed healthy young people. During the 1957 pandemic more than 50 percent of infections occurred in children and teenagers, at least a thousand of whom died of the disease.

When it was over, the influenza pandemic of 1957 had killed seventy thousand Americans and four million people worldwide. But Hilleman's quick actions saved thousands of American lives. The surgeon general of the United States, Leonard Burney, said, "many millions of persons, we can be certain, did not contract Asian flu because of the protection of the vaccine." For his efforts, Maurice Hilleman won the Distinguished Service Medal from the American military. "On that particular day," recalled Hilleman, "I was told to appear at the White House at 10 a.m. and bring my wife—and to wear a necktie, for God's sake."

Hilleman's committee-of-one approach would be hard to duplicate today. No American-based companies make inactivated influenza vaccine; the three companies that provide vaccine to the United States are headquartered in France, Switzerland, and Belgium. And regulatory control of vaccines by the U. S. Food and Drug Administration (FDA) would be impossible to ignore.

From 2003 to 2005, during the last few years of his life, Maurice Hilleman watched as bird flu spread from Hong Kong outward, eerily reminiscent of what he had seen in 1957. He also watched as bird flu spread from chickens to small mammals to large mammals to people. But months before his death Maurice Hilleman—the only man to accurately predict an influenza pandemic—made one more prediction, this one regarding the next pandemic. Understanding his prediction depends on understanding several aspects of influenza biology.

Influenza got its name from Italian astrologers, who blamed the disease's periodic occurrence on the influence—in Italian, *influenza*—of the heavenly bodies. The most important protein of influenza virus is the hemagglutinin, which attaches the virus to cells that line the windpipe and lungs. Antibodies to the hemagglutinin prevent influenza virus from binding to cells and infecting them. But influenza virus doesn't have only one type of hemagglutinin; it has sixteen.

Hilleman was the first to prove that these hemagglutinins change slightly every year. Because the hemagglutinin changes, Hilleman predicted in the early 1950s that protection against influenza would require yearly vaccination. "I could see that with time these viruses kept changing, changing, changing," recalled Hilleman, "and that this would explain why this virus could come back each year."

But sometimes influenza virus underwent a change so massive and so complete that no one in the population had antibodies to it. Such a change in the virus could cause a pandemic. Public health officials worry that the bird flu circulating in Southeast Asia since 2003 could be such a strain. Hilleman didn't think so. Bird flu is hemagglutinin type 5, or H5. Although H5 viruses can very rarely cause severe and fatal disease in people, the spread of H5 virus from one person to another is very poor. When Asian flu began in Hong Kong in 1957, the virus quickly spread from person to person, infecting about 10 percent of the population. When bird flu raged through chickens in Southeast Asia in 1997 and again in 2003, only three people caught the infection from another person; everyone else caught the disease directly from birds. Hilleman reasoned that bird flu wouldn't become a pandemic virus until it spread easily among people. H5 viruses have circulated for more than a hundred years and have never been very contagious. Hilleman believed that they never would be. He noted that only three types of hemagglutinins had ever caused pandemic disease in humans: H1, H2, and H3.

Hilleman believed that the future of influenza pandemics could be predicted from past pandemics:

H2 virus caused the pandemic of 1889.
H3 virus caused the pandemic of 1900.
H1 virus caused the pandemic of 1918.
H2 virus caused the pandemic of 1957.
H3 virus caused the pandemic of 1968.
H1 virus caused the mini-pandemic of 1986.

Hilleman saw two patterns in these outbreaks. First, the types of hemagglutinins occurred in order: H2, H3, H1, H2, H3, H1.

Second, the intervals between pandemics of the same type were always sixty-eight years—not approximately sixty-eight years, but exactly sixty-eight years. For example, an H3 pandemic occurred in 1900 and 1968, and an H2 pandemic occurred in 1889 and 1957. Sixty-eight years was just enough time for an entire generation of people to be born, grow up, and die. "This is the length of the contemporary human life span," said Hilleman. "A sixty-eight-year recurrence restriction, if real, would suggest that there may need to be a sufficient subsidence of host immunity before a past virus can regain access and become established as a new human influenza virus in the population." Using this logic, Hilleman predicted that an H2 virus, similar to the ones that had caused disease in 1889 and 1957, would cause the next pandemic—a pandemic that would begin in 2025. Only partially tongue-in-cheek, he declared his prediction to be "of greater predictive reliability than either the writings of Nostradamus or the *Farmers' Almanac*." When Hilleman made his prediction in 2005, knowing that his death was imminent, he said, "We'll soon know whether I was right or wrong. And I'll be watching. I'll be looking down—or up—to find out what happened." (Hilleman's prediction didn't hold up. The next influenza pandemic wasn't caused by an H2 strain in 2025; it was caused by an H1 strain in 2009.)

# CHAPTER 2

# Jeryl Lynn

*"I guess I'm famous because I got the mumps."*

JERYL LYNN HILLEMAN

Following the 1957 influenza pandemic, Maurice Hilleman left Walter Reed to become the director of virus and cell biology at Merck Research Laboratories. His goal at Merck—to prevent every viral and bacterial disease that commonly hurt or killed children—was unattainable. For the next three decades Hilleman made and tested more than twenty vaccines. Not all of them worked. But Hilleman came remarkably close to reaching his goal.

ONE VACCINE ORIGINATED FROM THE BACK OF HIS DAUGHTER'S THROAT. On March 23, 1963, at 1:00 a.m., Jeryl Lynn Hilleman woke up with a sore throat. Five years old, with penetrating blue eyes and an adorable pixie haircut, Jeryl quietly tiptoed into her father's bedroom and stood at the foot of his bed. "Daddy?" she whispered. Hilleman shook himself awake, rose to his full height of six feet one inch, bent down, and gently touched the side of his daughter's face. There, at the angle of her jaw, he felt a lump. Jeryl winced in pain.

At the time of his daughter's illness, Hilleman was a single father. Four months earlier his wife, Thelma, had died of breast cancer. "[Thelma] was the best-looking girl in Custer County High School," remembered Hilleman. "We were married in Miles City on New Year's Eve, 1944. For our honeymoon, we rode the train from Miles City back to the University of Chicago. When she got breast cancer, it went through her like wildfire. The chemotherapy in those days was crude. I worked during the day and would spend all night with Thelma at the hospital in Philadelphia."

Although he wasn't sure what was happening to Jeryl, Hilleman had a pretty good idea. Near his bed was a book titled *The Merck Manual*, a simply written compendium of medical information. Thumbing through it, he soon found what he was looking for. "Oh my God," he said, "you've got the mumps." Then Hilleman did something that few fathers would have done. He walked down the hallway, knocked on the housekeeper's door, and told her that he would be gone for a while. Then he went back to his bedroom, picked up his daughter, and put her back to bed. "I'll be back in about an hour," he said. "Where are you going, Daddy?" asked Jeryl. "To work, but I won't be long." Hilleman got into his car and drove fifteen miles to Merck. He rummaged around his laboratory, opening and closing drawers, until he found cotton swabs and a vial of straw-colored nutrient broth. By the time he got home, Jeryl had fallen back to sleep. So he gently touched her shoulder, woke her up, stroked the back of her throat with the cotton swab, and inserted it into the vial of broth. Then he comforted her, drove back to work, put the nutrient broth in a laboratory freezer, and drove home.

Most parents thought that mumps was a mild, short-lived illness. But Hilleman knew better. He was scared about what might happen to his daughter.

IN THE 1960S, MUMPS VIRUS INFECTED A MILLION PEOPLE IN THE United States every year. Typically the virus attacked the glands just in front of the ears, causing children to look like chipmunks. But sometimes the virus also infected the lining of the brain and spinal cord, causing meningitis, seizures, paralysis, and deafness. The virus didn't stop there. It also infected men's testes, causing sterility, and pregnant women, causing birth defects and fetal death. And it attacked the pancreas, causing diabetes. Although he knew that it was too late for Jeryl, Hilleman wanted to find a way to prevent mumps. He decided to use his daughter's virus to do it.

As with his work on influenza virus, Hilleman turned to chickens. When he got back to his laboratory, he took the broth containing Jeryl's virus and inoculated it into an incubating hen's egg; in the center of the egg was an unborn chick. During the next few days the virus grew in the membrane that surrounded the chick embryo. Hilleman then removed the virus and inoculated it into another egg. After passing the virus through several different eggs, Hilleman tried something else. He took an egg that had been incubating for twelve days and removed the gelatinous, dark brown chick embryo. Typically it takes about three weeks for an egg to hatch, so the embryo was still very small, weighing about the same as a teaspoon of salt. Hilleman cut off the head of the unborn chick, minced the body with scissors, treated the fragments with a powerful enzyme, watched the chick embryo dissolve into a slurry of individual cells, and placed the cells into a laboratory flask. (A cell is the smallest unit in the body capable of functioning independently. Human organs are composed of billions of cells.) Chick cells soon reproduced to cover the bottom of the flask. Hilleman passed Jeryl Lynn's mumps virus from one flask of chick cells to the next and watched as the virus got better and better at destroying the cells. He repeated this procedure five times.

Hilleman reasoned that as his daughter's virus adapted to growing in chick cells, it would get worse at growing in human cells. In other words, he was trying to weaken his daughter's virus. He hoped that the weakened mumps virus would then grow well enough in children to induce protective immunity, but not so well that it would cause the disease. When he thought the virus was weak enough, he turned to two friends, Robert Weibel and Joseph Stokes Jr., for help. Weibel was a pediatrician who worked in the Havertown section of Philadelphia, and Stokes was the chairman of pediatrics at the Chil-

*Maurice Hilleman and coworkers Joseph Stokes Jr. (left) and Robert Weibel, circa mid-1960s.*

dren's Hospital of Philadelphia. Stokes, Weibel, and Hilleman then made a choice that was typical of the time but abhorrent today: they decided to test their experimental mumps vaccine in developmentally disabled children.

In the 1930s, 1940s, 1950s, and 1960s, scientists often tested their vaccines in developmentally disabled children. Jonas Salk tested early preparations of his polio vaccine in disabled children at the Polk State School outside of Pittsburgh. At the time of Salk's experiments, no one in the government, the public, or the media objected to such testing. Everyone did it. Hilary Koprowski, working for the pharmaceutical company Lederle Laboratories, put his experimental live polio vaccine into chocolate milk and fed it to several disabled children in Petaluma, California, and a research team at Boston Children's Hospital used disabled children to test an experimental measles vaccine.

Today we see studies on developmentally disabled children as monstrous. We assume that scientists viewed disabled children as expendable subjects for their experimental, potentially dangerous vaccines. But disabled children weren't the only ones first subjected to

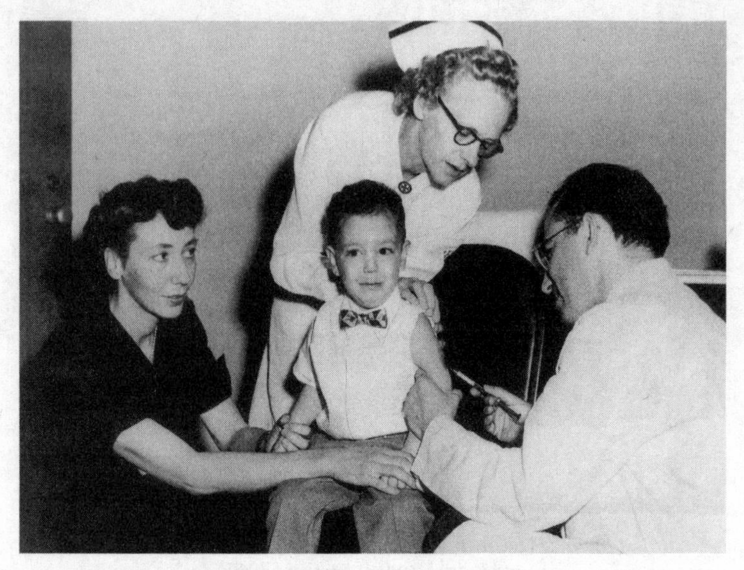

*Jonas Salk inoculates son Jonathan: May 16, 1953 (courtesy of the March of Dimes Birth Defects Foundation).*

these vaccines; researchers also used their own children. In July 1934, one year before he learned that his ill-fated polio vaccine inadvertently paralyzed and killed children, John Kolmer, working at Temple University in Philadelphia, inoculated his fifteen- and eleven-year-old sons with it. In the spring of 1953, Jonas Salk injected himself, his wife, and his three young children with an experimental polio vaccine—the same vaccine that he had given to developmentally disabled children at the Polk State School.

Salk believed in his vaccine, and he wanted his children to be among the first to be protected. "It is courage based on confidence, not daring," he said. "Kids were lined up in the kitchen to get the vaccine," recalled his wife, Donna. "I remember taking it for granted. I had complete and utter confidence in Jonas."

And one of the first children to receive Hilleman's experimental mumps vaccine was his second daughter, Kirsten. (Hilleman had remarried late in 1963.)

Why would researchers inject developmentally disabled children at the same time that they injected their own children with experimental vaccines? How can one reconcile these two apparently irreconcilable facts? The answer is that disabled children—confined to institutions where hygiene was poor, care was negligent, and space was inadequate—were at greater risk of catching contagious diseases, and of dying of those diseases, than other children. Disabled children living in large group homes suffered severe and occasionally fatal infectious diseases more commonly than other children. They weren't tested because they were more expendable; they were tested because they were more vulnerable.

Years later, Maurice Hilleman was unrepentant about his choice to test his mumps vaccine in disabled children. "Most children, [developmentally disabled] or not, get the mumps," he said. "My vaccine gave all of these children the chance to avoid the harm of that disease. Why should [developmentally disabled] children be denied that chance? [Developmentally disabled] children are perceived as more helpless; but their understanding of the shot that they were given, and their willingness to participate in the trial, was no different than [that of] healthy babies and young children. The difference was that [developmentally disabled] children were often wards of

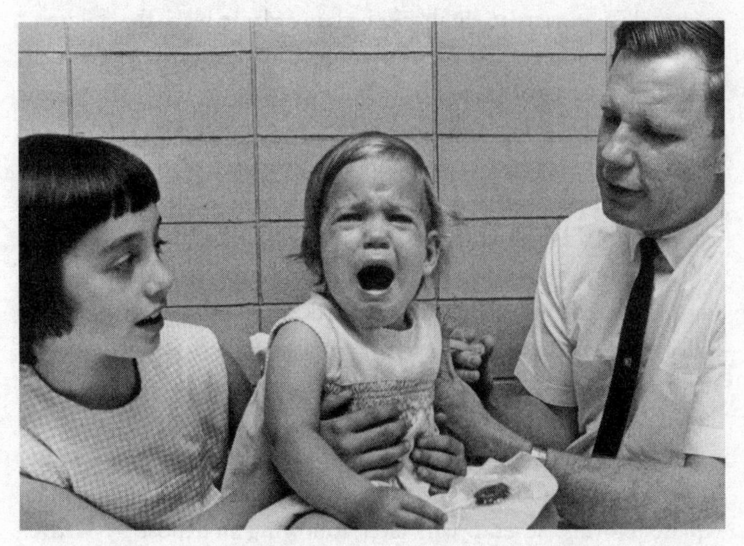

*Robert Weibel injects Kirsten Hilleman with the Jeryl Lynn strain of mumps vaccine, 1966. Jeryl Lynn Hilleman looks on.*

the state; their rights weren't protected by their parents. And although I think that states generally did a good job of protecting those rights, there have been abuses. And those abuses eventually changed how we thought about studying developmentally disabled children." Hilleman was talking about Willowbrook.

The Willowbrook State School was founded in 1938, when the New York state legislature purchased 375 acres of land on Staten Island and authorized the building of a facility for the care of developmentally disabled children. Construction was completed in 1942. The residents of Willowbrook were the most severely disabled, the most handicapped, and the most helpless of those being cared for in the New York state system. Although Willowbrook was designed to house three thousand people, by the mid–1950s about five thousand lived there. Jack Hammond, the director of Willowbrook, described the hellish, medieval living conditions: "When the patients are up and in the day rooms, they are crowded together, soiling, attacking each other, abusing themselves and destroying their clothing. At night, in many of the dormitories, the beds must be placed together in order to provide sufficient space for all patients. Therefore, except for one narrow aisle, it is virtually necessary to climb over beds in order to reach the children."

A small, poorly trained staff coupled with massive overcrowding led to a series of tragedies at Willowbrook. In 1965, inadequate supervision by a teenaged attendant caused a forty-two-year-old resident to be scalded to death in a shower. A few months later, a ten-year-old boy suffered the same fate in the same shower. That same year, a twelve-year-old boy died of suffocation when a restraining device loosened and twisted around his neck. The following month, one of the residents struck another in the throat and killed him. At the end of the year Senator Robert Kennedy paid a surprise visit to Willowbrook. Horrified by what he saw, Kennedy called Willowbrook "a new snakepit" and said that the facilities were "less comfortable and cheerful than the cages in which we put animals in the zoo." Kennedy's visit prompted short-lived, inadequate reforms. Several years later, following an exposé by WABC-TV in New York titled "Willowbrook: The Last Great Disgrace," legislators finally instituted meaningful reforms. The reporter for the story was a twenty-nine-year-old journalist recently hired by the station: Geraldo Rivera.

In addition to abuse and neglect, the overcrowding, poor sanitation, and inadequate staff at Willowbrook led to the spread of many infectious diseases, including measles, influenza, and shigellosis, and those caused by intestinal parasites. But no infection was more damaging than hepatitis. In an effort to control outbreaks of hepatitis, the medical staff at Willowbrook consulted Saul Krugman, an infectious disease specialist at Bellevue Hospital in New York City. Krugman found that hepatitis developed in 90 percent of children admitted to Willowbrook soon after their arrival. Although it was known that hepatitis was caused by a virus, it wasn't known how hepatitis virus spread, whether it could be prevented, or how many different types of viruses caused the disease. Krugman used the children of Willowbrook to answer those questions. One of his studies involved feeding live hepatitis virus to sixty healthy children. Krugman watched as their skin and eyes turned yellow and their livers grew bigger. He watched them vomit and refuse to eat. All the children fed hepatitis virus became ill, some severely. Krugman reasoned that it was justifiable to inoculate disabled children at Willowbrook with hepatitis virus because most of them would get hepatitis anyway. But by purposefully giving the children hepatitis, Krugman increased that chance to 100 percent. "They were the most unethical medical experiments ever performed in children in the United States," said Hilleman. Art Caplan, director of the Center for Bioethics at the University of Pennsylvania, agrees. "The Willowbrook studies were a turning point in how we thought about medical experiments in [developmentally disabled] children," said Caplan. "Children inoculated with hepatitis virus had no chance to benefit from the procedure—only the chance to be harmed."

Caplan believes that the studies at Willowbrook weren't the only reason for the change in public sentiment about experiments on disabled children. In the early 1950s, scientists at the Massachusetts Institute of Technology (MIT) were interested in determining how people absorbed iron, calcium, and other minerals from food. The scientists went to the Walter E. Fernald School, an institution for mentally or physically disabled children in Waltham, Massachusetts, about twelve miles outside of Boston. Conditions at Fernald weren't nearly as barbaric as those in Willowbrook. Unlike Willowbrook, the Fernald School had a science club for teenagers. Members of this club would eventually participate in an experiment that put an end to studies on disabled children.

Funded by the National Institutes of Health (NIH), the Atomic Energy Commission, and the Quaker Oats Company, MIT researchers fed children breakfast foods tagged with small amounts of radioactive cobalt. They wanted to determine whether minerals contained in Quaker Oats products would move through the body in a manner different from, and presumably better than, minerals in other breakfast foods. Although the amount of radiation exposure was small—about three hundred millirems, less than the yearly radiation exposure of someone living at high altitudes, such as that of Denver—children given the radioactive food had no chance of benefiting from the experiment.

Today, because of the studies at Willowbrook and Fernald, medical researchers don't include disabled children in studies from which they cannot benefit. They also don't include them in studies from which they might benefit.

On June 28, 1965, about two years after Jeryl Lynn Hilleman had walked into her father's bedroom, Robert Weibel visited the Trendler School. Located in an old converted two-story house in Bristol, Pennsylvania, the school was home to thirty severely disabled children. Weibel injected sixteen of them with Hilleman's experimental mumps vaccine. (For Weibel the decision to inoculate disabled children had special significance. Weibel's son, Robert, had been born with Down syndrome in 1956.) Weibel found that Hilleman's vaccine was safe and that antibodies against mumps developed in vaccinated children. Encouraged, Weibel went to the Merna Owens and St. Joseph's Homes, both isolated in the countryside of northeast Pennsylvania. He found sixty severely disabled children who were susceptible to mumps, and on August 13, 1965, he injected half of them with Hilleman's vaccine. The results were the same. Children developed mumps antibodies but didn't get sick.

Hilleman, Weibel, and Stokes had shown that their vaccine induced mumps antibodies, but they hadn't proved that it prevented disease. To do that, they needed more children. For the next several months, Stokes and Weibel handed out fliers describing the new mumps vaccine to nursery schools and kindergartens in the Philadelphia area. If parents were interested, they could attend informational meetings to be held at local churches. "Parents were eager to join up," said Weibel. "They trusted us, and we were straightforward with them. They knew that diseases like measles and mumps could be pretty bad, and they wanted to help out. In those days people were a little less focused on themselves and more focused on their community. They wanted to make a difference."

Hilleman attended two of the meetings in the Philadelphia suburb of Havertown, but he didn't do any of the talking. "At the Sacred Heart Church Maurice wanted to remain inconspicuous," recalled Weibel, "so he stood in the back. He was leaning up against the door frame when he accidentally backed into a vessel of holy water nailed to the wall. He drenched the back of his shirt and was quite shaken, apologizing endlessly to the priest. The priest reassured him by saying that there was plenty where that came from." During another meeting, Hilleman remembered a parent asking how the vaccine was made. Joe Stokes answered the question. Handsome, with silver-gray hair and the gentle spirit of his Quaker background, Stokes told the story of a German man who had left his wedding ring on a nightstand. During the night a spider made a dense, intricate web by going back and forth, side to side. By the morning the web had covered the entire opening in the ring. "He was trying to explain what chick cells looked like when they grew in laboratory flasks," recalled Hilleman. "He called it *Gewebekulter*—web culture. Jesus, they were spellbound."

Parents interested in participating in the study received a three-by-five-inch card stating, "I allow my child to get a mumps vaccine." At the bottom of the card was a line for the parents' signatures. Unlike the practice today, the consent card didn't contain an explanation of the disease, a description of the vaccine, a list of vaccine components, a discussion of previous studies, the need for blood tests, or a statement of possible risks and benefits. Parents were also given Robert Weibel's work and home telephone numbers. If they had any questions about the vaccine or if they were worried that the vaccine was causing a problem, they could call him at any time, day or night. Weibel, in turn, would drive to their homes and examine their children.

Stokes and Weibel recruited about four hundred children for their study; two hundred received Hilleman's vaccine, and two hundred received nothing. "Then we waited," recalled Weibel. Several months later, a mumps epidemic swept through Philadelphia. Sixty-three children in the study came down with mumps. Two of the sixty-three had been vaccinated. Sixty-one of them had not been. Hilleman's vaccine worked, and it worked well. On March 30, 1967, four years after Jeryl Lynn Hilleman had come down with the mumps, the Jeryl Lynn strain of mumps vaccine was licensed. Since then, more than three hundred million doses have been distributed in the United States. By 2000, mumps vaccine had prevented almost one million children from getting mumps every year and had prevented meningitis and deafness in thousands. Furthermore, the introduction of Hilleman's mumps vaccine in Denmark, Finland, Norway, Sweden, Slovenia, Croatia, England, Wales, Israel, Poland, Romania, and Latvia has virtually eliminated the disease from those countries.

Jeryl Lynn Hilleman is now the executive vice president and chief financial officer of Symyx, a research technology company in Santa Clara, California. "People ask me what it means to be Jeryl Lynn, the namesake of the vaccine. I tell them that it just made me very proud of my father. If being sick at the right time with the right virus helped—great." One reporter later wrote, "Jeryl recovered from mumps virus, but mumps virus never recovered from infecting Jeryl."

# CHAPTER 3

# Eight Doors

*"If I have seen farther than others, it was because I was standing on the shoulders of giants."*

Sir Isaac Newton

What possessed Maurice Hilleman to take his daughter's mumps virus and inject it into hen's eggs and minced chick embryos? Why did he cut off the chicks' heads before using them? And most importantly, why in the 1960s did he resort to a process so seemingly crude, arcane, and convoluted? Eight critical experiments performed during the previous century determined Hilleman's choices.

From Edward Jenner, Hilleman learned the power of vaccines. Jenner's vaccine eradicated humankind's deadliest infection, smallpox, from the face of the earth.

Easily spread by tiny droplets of saliva containing millions of virus particles, smallpox was a common, severe, debilitating infection. The virus caused high fever and a permanently disfiguring, pus-filled rash with a smell reminiscent of rotting flesh. Smallpox killed one of every three of its victims and blinded many survivors. In 1492, when Christopher Columbus crossed the Atlantic Ocean, seventy-two million Indians lived in North America; by 1800, only six hundred thousand remained. Smallpox—brought by European settlers—killed most of the rest. Indeed, smallpox has killed more people than all other infectious diseases combined.

In 1768, when Edward Jenner was thirteen years old and training as an apprentice apothecary in Chipping Sodbury, England, he approached a young milkmaid who appeared ill. "Are you coming down with the smallpox?" he asked. "I cannot take that disease," she said, "for I have had the cowpox." Cowpox was a disease that caused blisters on the udders of cows. Sometimes people who milked cows with cowpox would get these same blisters on their hands. Jenner was only a boy, so he didn't give much thought to the milkmaid's notion of what prevented diseases. But Edward Jenner remembered that conversation for the rest of his life.

Years later, while training in London, Jenner told the famous surgeon John Hunter about the milkmaid's observation. Hunter encouraged Jenner to test the theory. "Don't think, but try," said Hunter. "Be patient, be accurate." On May 14, 1796, several months before George Washington gave his farewell address, Edward Jenner got his chance. Sarah Nelmes, a milkmaid in Jenner's employ, had cowpox blisters on her hands and wrists. Jenner removed the pus from one of the blisters and injected it into the arm of James Phipps, the eight-year-old son of a local laborer. Six weeks later, Jenner injected Phipps with pus taken from a case of smallpox "in order to ascertain whether the boy, after feeling so slight an affection of the system from the cowpox virus, was secure from contagion of smallpox." Typically, inoculation with smallpox caused high fever; chills; an ulcerating, painful rash; and occasionally death. But nothing happened to James Phipps. Later, Jenner injected Phipps twenty more times with pus from people with smallpox; each time Phipps survived without incident. Apparently, cowpox virus was similar enough to smallpox so that inoculation with one protected against disease caused by the other.

Two years later, Jenner published his observations under the lengthy title "An Inquiry into the Causes and Effects of Variolae Vaccinae, a Disease Discovered in Some of the Western Counties of

England, Particularly Gloucestershire, and Known by the Name of Cow Pox." Jenner used the term *variolae vaccinae*—literally "small-pox of the cow" and later the source of the word *vaccine*. Within one year of his publication, physicians had inoculated a thousand people with cowpox and had translated Jenner's observations into several languages. It took about two hundred years for Jenner's vaccine to eradicate smallpox from the face of the earth. (Although the disease is gone, the virus isn't. Fearing that smallpox virus—secretly preserved in scientific laboratories—would be used as a weapon of terror, the United States government supported a short-lived program to immunize hospital workers in October 2002, five months before the invasion of Iraq.)

Despite Jenner's success, scientific advances often come with a price—the landscape of vaccines is littered with tragedy. Jenner lacked a reliable, consistent, and continual source of cowpox virus. So he inoculated cowpox under the skin of a volunteer, waited eight days until it caused a blister, removed the pus, and inoculated it into the arm of the next person. Many children were vaccinated by this arm-to-arm technique. For example, in St. Petersburg, Russia, in 1801 a recently vaccinated girl was sent to a local orphanage to serve as a source of cowpox virus for other children. The orphanage continued arm-to-arm inoculation for more than ninety years. But arm-to-arm transfer of cowpox could be dangerous. One child inoculated by Jenner, a five-year-old boy named John Baker, was never challenged with smallpox. "The boy," said Jenner, "was rendered unfit for inoculation [with smallpox] from having felt the effects of a contagious fever in a workhouse soon after this experiment was made." Baker was unfit for inoculation because he had died of a bacterial infection, probably the result of a contaminated cowpox vaccine. In 1861 in Italy forty-one children got syphilis as a result of arm-to-arm transfer when a small amount of blood from one child in the chain, who had an undiagnosed case of the disease, was injected into others. And in 1883 in Bremen, Germany, arm-to-arm transfer caused a massive outbreak of hepatitis.

Although Edward Jenner made the first viral vaccine, he didn't know that smallpox and cowpox were related viruses. That was because he'd never heard of viruses. Edward Jenner made his observations several decades before scientists showed what viruses were and how they reproduced.

From Louis Pasteur, a French chemist, Hilleman learned that vaccines could be made from dangerous human viruses. (Jenner had used a cow virus.) Pasteur developed humankind's second vaccine, one that prevented a uniformly fatal disease: rabies.

On July 4, 1885, a rabid dog attacked a nine-year-old boy named Joseph Meister in the town of Meissengott, a small village in the province of Alsace, France. Meister, who was on his way to school, covered his face as the dog knocked him down and bit him fourteen times. A bricklayer walking nearby beat the dog with an iron bar and carried Meister home. The owner later killed the dog and cut open its stomach; out poured straw, hay, and fragments of wood—evidence that the animal had gone mad. (Old stories about infectious diseases often sound as if they had been written by the Brothers Grimm.)

In ancient times, people with rabies were hunted down like wild animals and strangled or suffocated. By the late 1800s, treatments for rabies had advanced to include the feeding of cock's brains, crayfish eyes, livers from mad dogs, snake skins mixed with wine, and poison from a viper, or the "dipping cure," which involved holding victims under water until "they have done the kicking." Techniques that actually worked to prevent rabies included immediately cauterizing a bite with a hot iron or sprinkling gunpowder on the wound and igniting it; these processes killed the virus.

Two days after the attack, Joseph Meister and his mother arrived at the front door of 45 rue d'Ulm in Paris, the home of Pasteur's laboratory. When Pasteur came to the door, Meister's mother dropped to her knees and begged him to save her son. Pasteur took the boy by the hand and gently guided him into his home, later describing the wounded child in his notebook, "Severely bitten on the middle finger of his right hand, on the thighs, and on the leg by the same rabid dog that tore his trousers, threw him down and would have devoured him if it had not been for the arrival of a mason armed with two iron bars who beat down the dog."

For several years preceding Meister's visit to his laboratory, Pasteur had studied rabies virus. To make an experimental rabies vaccine, he found dogs that had died of rabies, ground up their spinal cords, injected infected spinal cords into rabbits, and watched the rabbits die of rabies. Then he removed the rabbits' spinal cords, cut them into thin strips, and dried them in airtight jars. Pasteur found that the longer he dried them, the longer it took for the infected spinal cords to cause disease. After fifteen days of drying, they didn't cause disease at all. Apparently, prolonged drying killed rabies virus. Pasteur then performed his groundbreaking experiment. He injected dogs with rabies-infected spinal cords that had been dried for fifteen days and then, successively, with spinal cords that had been dried for fewer and fewer days. At the end of the experiment, Pasteur injected dogs with spinal cords that contained live, deadly rabies virus. Typically, the dogs would have died of rabies. But all the dogs that received Pasteur's vaccine survived.

When Joseph Meister came to his laboratory, Pasteur had not yet immunized people, only animals. But at 8:00 p.m. on July 6, 1885, Meister was injected with a rabies-infected rabbit spinal cord that had been dried for fifteen days. Pasteur knew that such a spinal cord didn't kill dogs or rabbits. He could only hope that it wouldn't kill Meister. During the next eleven days, Meister was injected twelve more times with rabbit spinal cords that had each been less and less dried out and therefore were more and more likely to cause rabies. The final dose, on July 16, was taken from an infected rabbit spinal cord that had been dried for only one day—an injection that would have easily killed a rabbit. Pasteur knew that those final injections were potentially deadly. Writing to his children, he said, "this will be another bad night for your father. [I] cannot come to terms with the idea of applying a measure of last resort to this child. And yet [I have] to go through with it. The little fellow continues to feel very well."

By the end of the month, Meister was home in Alsace, healthy. Using killed, partially killed, and live rabies virus, Pasteur had developed the first vaccine that protected people bitten by rabid animals from getting rabies. Parisians, who had to live every day in fear of rabid dogs prowling their streets, hailed Pasteur's vaccine as one of the greatest medical triumphs of the nineteenth century. But like Jenner's smallpox vaccine, Pasteur's rabies vaccine came with a price. As his vaccine was injected into more and more people, Pasteur found something that he hadn't anticipated: some people—as many as one of every two hundred who used it—became paralyzed and died. At first, Pasteur thought that people were dying of rabies. But they were dying of a reaction to his vaccine.

Today we understand the problem with Louis Pasteur's rabies vaccine. Cells from the brain and spinal cord contain a substance called myelin basic protein. This protein forms a sheath around nerves, like the rubber insulation that surrounds an electrical wire. Some people inoculated with myelin basic protein occasionally have an immune response against their own nervous systems: autoimmunity. Pasteur's vaccine, made from rabbit spinal cords that contained myelin basic protein, caused autoimmunity. (This was why Hilleman cut off the heads of chick embryos before using them. He didn't want to inject children with small amounts of myelin basic protein from the chicks' brains.)

Joseph Meister, who survived the bite of a rabid animal, lived to be sixty years old. When the Nazis occupying Paris in 1940 wanted to see the tomb of Louis Pasteur, Meister, then a guard at the Pasteur Institute, was the first to meet them. But the humiliation of opening his savior's tomb to the Nazi invaders was more than he could handle. Later, locking himself in his small apartment, Meister committed suicide.

FROM MARTINUS BEIJERINCK, A PROFESSOR OF BACTERIOLOGY AT THE Delft Polytechnic Institute in the Netherlands, Hilleman learned what viruses were, where they reproduced, and how they caused disease.

As Peter Radetsky describes in *The Invisible Invaders*, Beijerinck "would burst into his lab, a tall, striking figure in a dark coat and high collar. Around the rooms he would prowl, shutting all windows, disdainfully sniffing for the faintest remnant of cigarette smoke, and inspecting benches for as little as a drop of spilled water." A mean-spirited, haughty, offensive man, Beijerinck often likened his students to untrained monkeys and refused to allow young associates to marry. His personality didn't limit his achievements, however. In 1898, Martinus Beijerinck performed an experiment that revolutionized microbiology.

Beijerinck was studying tobacco mosaic disease, which stunted the growth of tobacco plants and was common in Europe and Russia. Scientists had already seen bacteria under the microscope, shown that they caused specific diseases, and figured out a method to remove them from water: filtration through unglazed porcelain. (Pots of unglazed porcelain were often kept in the home to purify drinking water.) Beijerinck assumed that bacteria caused tobacco mosaic disease. To prove it, he squeezed diseased plants through a press, collected the sap, rubbed the sap onto healthy leaves, and watched the healthy plants die. Clearly, the sap contained the organism that caused the disease. Then Beijerinck performed his seminal experiment. He passed infectious sap through a porcelain filter and, much to his surprise, found that the sap still caused disease. Beijerinck knew that bacteria should have been trapped by the filter. Something else was getting through.

Beijerinck published his findings in a paper titled "Concerning a Contagium Vivum Fluidum as a Cause of the Spot-Disease of Tobacco Leaves." The term *contagium vivum fluidum* translates as "living contagious fluid." (Later, Beijerinck referred to the *contagium* as a virus.) Beijerinck said that "the contagium, in order to reproduce, must be incorporated into the living protoplasm of the cell." Martinus Beijerinck had recognized the single most important difference between bacteria and viruses. Bacteria, capable of independent growth, can multiply on the surface of furniture, in dust, in rainwater, or on the lining of the skin, nose, or throat. But viruses, incapable of independent growth, can reproduce only within the "living protoplasm of the cell." At the age of forty-seven, Martinus Beijerinck became the father of virology.

. . . .

JENNER NEEDED COWS TO MAKE HIS SMALLPOX VACCINE; PASTEUR needed dogs and rabbits. From Alexis Carrel, Hilleman learned that animal organs could be kept alive outside of the body, freeing researchers from using whole animals when they wanted to make their vaccines.

On January 17, 1912—three months before the steamship *Titanic* sank in the Atlantic Ocean—Carrel, a French surgeon working at the Rockefeller Institute in New York City, removed a small piece of heart from an unhatched chick embryo and placed it in the bottom of a flask. Every two days he added nutrient fluid that contained chicken plasma and a crude extract made from a chicken embryo. He wanted to see how long he could keep the piece of chicken heart alive. Obsessed that the heart might be inadvertently contaminated with bacteria, Carrel created a cult around its maintenance, insisting that the walls be painted black and that his technicians wear long black gowns with hoods when they entered the room in which it was kept. To celebrate their success, every January doctors and nurses at the Rockefeller Institute lined up outside the research laboratory, locked hands, and joined Carrel in lustily singing "Happy Birthday" to the small piece of heart. Carrel and his colleagues maintained the chicken heart fragment until his death in 1944.

FROM ERNEST GOODPASTURE, WHO WORKED IN THE EARLY 1930S, Hilleman learned that viruses could be grown in eggs, a discovery that forged a permanent bond between virologists and chicken farmers.

Born on a farm near Clarksville, Tennessee, Goodpasture, a quiet, unassuming pathologist, was interested in fowlpox, a virus similar to smallpox. Because fowlpox infected chickens, he decided to try to grow the virus in hens' eggs, reasoning that eggs were sterile (antibiotics hadn't been invented yet) and inexpensive. Working at Vanderbilt University in Nashville, Goodpasture took an incubating hen's egg, bathed it in alcohol, and, to sterilize the shell, set it on fire. Then, using an eggcup as an operating table, he cut a small window in the shell and injected the egg with fowlpox. The virus grew readily in the membrane surrounding the chick embryo. Hilleman used Goodpasture's technique to make his pandemic influenza and mumps vaccines.

FROM MAX THEILER, HILLEMAN LEARNED THAT HUMAN VIRUSES COULD be weakened and made into vaccines by growing them in animal cells. (Remember, Hilleman weakened his daughter's mumps virus by growing it in chick cells.)

Theiler, a South African émigré also working at the Rockefeller Institute, wanted to make a vaccine to prevent yellow fever, a tropical viral disease that caused bleeding, the unmistakable symptom of black vomit, jaundice—a yellowing of the eyes and skin that gave the virus its name—and death. Because yellow fever virus caused severe internal bleeding, it was called a viral hemorrhagic fever. Yellow fever was common in the United States; an outbreak in Philadelphia in the late 1700s killed 10 percent of the city's residents, and an outbreak in New Orleans in the mid-1800s killed 30 percent. The terror once caused by yellow fever is associated today with another viral hemorrhagic fever: Ebola virus.

In the mid-1930s, Max Theiler performed a series of experiments that determined how researchers would make viral vaccines for the next ninety years. Using Carrel's technique of growing chopped-up animal organs in laboratory flasks, Theiler found that yellow fever virus grew in mouse embryos. So he passed the virus from one mouse embryo to another and eventually from one chicken embryo to another. Theiler reasoned that as yellow fever virus got better and better at growing in cells from different species—like mice and chickens—it would become less and less capable of causing disease in humans. (Today we know that human viruses forced to grow in animal cells undergo a series of genetic changes that make them less capable of reproducing and causing disease in people.) To test his theory, Theiler injected a thousand Brazilians with what he hoped was a weakened form of yellow fever virus. He found that most people developed antibodies to the virus and that no one got the disease. By the end of the 1930s, Theiler had inoculated more than half a million Brazilians, and epidemics of yellow fever in Brazil abated. The yellow fever vaccine made in mouse embryos by Max Theiler in the mid-1930s is still used today.

Theiler's technique of weakening human viruses by growing them in cells from other species remains the single most important method for making live weakened viral vaccines. His method has been used to make vaccines against measles, mumps, rubella, chickenpox, polio, and rotavirus. In 1951, "for his discoveries concerning yellow fever and how to contain it," Max Theiler won the Nobel Prize in medicine. When asked what he planned to do with the $36,000 in prize money, he said "Buy a case of Scotch and watch the [Brooklyn] Dodgers."

Like Jenner and Pasteur before him, Theiler also saw tragedy follow his vaccine. In the early 1940s, scientists made Theiler's vaccine using human serum obtained from several volunteers, to stabilize the virus. Unfortunately, unnoticed at the time, at least one of these volunteers was jaundiced, infected with hepatitis B virus. As a consequence, more than three hundred thousand American servicemen injected with contaminated yellow fever vaccine got hepatitis, and sixty died. Human serum was never again used to stabilize vaccines.

From the research team of John Enders, Thomas Weller, and Frederick Robbins, working at Boston Children's Hospital (part of Harvard's Medical School) in the late 1940s, Hilleman learned how to grow animal and human cells in the laboratory. Alexis Carrel's technique, using chopped up animal organs, was called tissue culture; the Enders group's technique, using single layers of animal or human cells grown in laboratory flasks, was called cell culture. Now when researchers want to grow viruses, they simply take a vial of cells out of the freezer, thaw them out, place them into laboratory flasks, watch them reproduce until a single layer neatly covers the bottom of the flask, and inoculate them with viruses. The days of growing viruses in whole animals or chopped-up animal organs were over. The Enders group's technique is still used to make viral vaccines today.

*John Enders (left) and Thomas Weller in their laboratory at Boston Children's Hospital, November 1954 (courtesy of the March of Dimes Birth Defects Foundation).*

One of the Boston group's first cell cultures was made from a human fetus. On March 30, 1948, at 8:30 a.m., Thomas Weller walked across the street in front of Boston Children's Hospital and into the office of Duncan Reid, an obstetrician working at the Boston Lying-In Hospital who had just aborted a twelve-week pregnancy. The mother had chosen to end her pregnancy because she had been infected with rubella, a virus known to cause birth defects. Reid handed Weller the fetus. After coaxing fetal cells to reproduce on the bottom of laboratory flasks, Weller found that polio virus grew in the cells. Weller, Robbins, and Enders found later that polio virus also grew in a variety of different animal and human cells. Prior to these experiments, polio virus could be grown only in cells from brains and spinal cords. Researchers feared using a polio vaccine made from nervous tissue for the same reason that they feared Pasteur's vaccine: the dangerous side effect of autoimmunity.

In 1954 Enders, Weller, and Robbins won the Nobel Prize in medicine for "their discovery of the ability of polio viruses to grow in cultures of various types of tissue." These studies allowed Jonas Salk and Albert Sabin to make polio vaccines that eventually eliminated polio from most of the world.

From Jonas Salk, the scientist who first found a way to prevent polio, Hilleman learned that vaccines could win the heart of the American public.

Born and raised in New York City, the son of Russian immigrants, Salk was driven, obstinate, and self-assured. Working at the University of Pittsburgh in the early 1950s, Salk used the Enders group's technique to grow polio virus in monkey kidney cells. Then he purified the virus, killed it with formaldehyde, and injected it into seven hundred children in and around Pittsburgh. Salk reasoned that killed polio virus would induce polio antibodies but wouldn't cause polio. In 1954, funded by the March of Dimes, doctors and nurses injected four hundred thousand children with Salk's vaccine and two hundred thousand with an inert liquid that looked like vaccine, called placebo. The program was then and remains today the largest test of a medical product ever performed. Following the announcement that the vaccine worked, Americans named hospitals, schools, streets, and babies after Salk and sent him money, clothes, and cars. Universities offered him honorary degrees, and countries issued proclamations in his honor. Salk started the day as a scientist at the University of Pittsburgh and ended it as one of the most revered men on the face of the earth. When people hear the word *vaccine* today, the first person they think of is Jonas Salk.

But like Jenner, Pasteur, and Theiler before him, Salk watched tragedy follow his vaccine. When Salk found that polio vaccine could be made by inactivating polio virus with formaldehyde, five companies stepped forward to make it. On April 12, 1955, each of those companies was permitted to sell its vaccine to the public. One company, Cutter Laboratories of Berkeley, California, made it badly. Researchers and executives at Cutter were confident that they had made their polio vaccine exactly as Jonas Salk had prescribed, giving it to the children of four hundred and fifty of their employees. But, because Cutter researchers hadn't properly filtered out the cells in which they grew polio virus, some virus particles had effectively escaped the killing effects of formaldehyde. As a consequence, more than one hundred thousand children were inadvertently injected with live, dangerous polio virus. Worse, children injected with Cutter's vaccine spread polio to others, starting the first and only man-made polio epidemic. When the dust settled, live polio virus contained in Cutter's vaccine had infected two hundred thousand people; caused about seventy thousand to have mild cases of polio; permanently and severely paralyzed two hundred people, mostly children; and killed ten. It was one of the worst biological disasters in American history. Federal regulators quickly identified the problem with Cutter's vaccine and established better standards for vaccine manufacture and safety testing. Cutter Laboratories never again made another polio vaccine. And Salk's polio vaccine helped to dramatically reduce—and in some countries eliminate—one of the world's most crippling infections.

HILLEMAN HAD LEARNED FROM WHAT HAD GONE RIGHT AND WHAT had gone wrong before him. By the time he made his mumps vaccine, an enormous path had been cleared through the thicket. "It was an age of genius," he said. "I was able to do what I did because of what they did."

# CHAPTER 4

## The Destroying Angel

*"That fatal and never to be forgotten year, 1759, when the Lord sent the destroying Angel to pass through this place, and removed many of our friends into eternity in a short space of time; and not a house exempt, not a family spared from the calamity. So dreadful was it that it made every ear tingle, and every heart bleed; in which time I and my family were exercised with that dreadful disorder, the measles. But by blessed God our lives were spared."*

Ephraim Harris, colonist and farmer, Fairfield, New Jersey

While Maurice Hilleman was making his mumps vaccine, he was also making a measles vaccine.

Measles infection starts innocently enough, with fever, cough, a runny nose, pinkeye, and rash. But measles virus also can infect the lungs, causing fatal pneumonia, and the brain, causing seizures, deafness, and permanent brain damage. And it can infect the liver, kidneys, heart, and eyes, blinding many survivors. Furthermore, measles virus causes one of the most insidious, unrelenting diseases of childhood—subacute sclerosing panencephalitis (SSPE), a rare but uniformly fatal disorder. Symptoms of SSPE usually begin about seven years after measles infection. At first children undergo subtle personality changes, their handwriting deteriorates, and they seem to forget things. Later, when the horror of the disease fully emerges, children are progressively less able to walk, stand, or talk; then they become combative, have seizures, lapse into a coma, and die. Despite decades of study, a parade of drugs, and heroic supportive measures, no child has ever survived SSPE.

In the early 1960s, when Maurice Hilleman wanted to make his vaccine, measles virus was killing eight million children in the world every year. Doctors and public health officials were desperate to find a way to prevent it.

The road to a measles vaccine started in Boston.

In 1954, Thomas Peebles was working in the laboratory of John Enders at Boston Children's Hospital. The team, which had just won the Nobel Prize in medicine for its work on polio, also included Sam Katz, a brilliant infectious diseases specialist and pediatrician from New Hampshire, and Milan Milovanovic, a scientist from Belgrade, Yugoslavia. Peebles, fresh from his internship at Massachusetts General Hospital, had gotten a late start in his career, spending four years in the navy during the Second World War. Enders assigned Peebles the task of capturing measles virus. Although researchers were sure that a virus caused measles, no one had ever coaxed it to reproduce inside a test tube.

In January 1954, Peebles got the break he was looking for. Dr. Theodore Ingalls called Peebles and told him about an outbreak of measles at the Fay School, an exclusive all-boys private school in Southborough, a suburb west of Boston. (The school, founded in 1866, was one of the oldest boarding schools in the United States.) Peebles got into his car, drove thirty miles, and convinced the school's principal, Harrison Reinke, to allow him to collect blood from the boys. Then he went to each student and said, "Young man, you are standing on the frontiers of science. We are trying to grow this virus for the first time. If we do, your name will go into our scientific report of the discovery. Now this will hurt a little. Are you game?"

For the next few weeks Peebles tried unsuccessfully to capture measles virus. But on February 8, 1954, his luck changed. David Edmonston, a thirteen-year-old student at Fay, had stomach cramps, nausea, fever, and a faint red rash that started on his face and spread to his chest, abdomen, and back. When David's temperature rose to 104 degrees, large quantities of measles virus coursed through his veins. Thomas Peebles was about to do something that no one had dared to do before: make a vaccine using human organs.

In the early 1950s, doctors subjected children to a medical procedure, now abandoned, for a condition called hydrocephalus—literally, "water on the brain." In the center of the brain, specialized cells make fluid that bathes the spinal cord; the fluid must travel through a series of narrow canals. Sometimes, because of infection or birth defects, those canals are blocked and spinal fluid is trapped, compressing the brain. To relieve the blockage, surgeons would drill a small hole into the center of the brain, insert a plastic catheter, and create a tunnel under the skin that ended in the ureter, a tube that connects the kidneys to the bladder. In order to connect the catheter to the ureter, surgeons needed to remove a kidney—one perfectly healthy, fully functioning human kidney. Now, instead of spinal fluid getting trapped in the brain, it was carried to the bladder and urinated out of the body.

Enders couldn't bear to see healthy kidneys wasted. "He was the typical penurious Yankee," recalled Sam Katz. "He just hated to see things wasted. And his thriftiness extended to [laboratory] materials." So Enders sent Thomas Peebles to collect the kidneys before they were thrown away and asked him to try to grow measles virus in them. First, Peebles treated the kidneys with a powerful enzyme to make sure the kidney cells didn't clump together. Then he put the cells into sterile flasks, watched them reproduce until they covered the bottom of each flask, and added David Edmonston's blood. After a few days, the kidney cells shriveled up and died. Edmonston's measles virus was apparently reproducing inside the cells, killing them in the process.

Now that the Boston researchers could grow measles virus in laboratory cells, they could weaken it to make a vaccine. Peebles, Katz, and Milovanovic grew the virus serially in twenty-four cultures of human kidneys, twenty-eight cultures of human placentas (Enders apparently couldn't stand to watch placentas getting thrown away either), six hens' eggs, and six minced chick embryos. They hoped that by forcing David Edmonston's measles virus to grow in this hodgepodge of human and animal cells, the virus would weaken sufficiently to be a vaccine. There was, however, no formula, no recipe, for how to do this.

Thomas Weller remembered what it was like to work in a laboratory dense with human kidneys and placentas: "I would walk across the street to the Boston Lying-In Hospital, and obstetricians would hand me these heavy placentas from babies that they'd just delivered. Some days there were no placentas. But on others there were as many as six. Back in Dr. Enders's laboratory I set up two ring stands and balanced a thick, sterile steel rod between them. Then I draped the placentas over the rods like clothes on a clothesline."

Now it was time for the Boston team to test the vaccine. So Sam Katz drove to the Fernald School, where a few years earlier researchers had fed radioactive cereal to members of its science club. "We chose the Fernald School because every year there were outbreaks of measles [in the school]," recalled Katz, "and every year several children died." On October 15, 1958, Katz injected eleven mentally or physically disabled children with the vaccine developed in Enders's laboratory. All the children developed protective antibodies. But eight of them had fever, and nine had a mild rash. Although the vaccine protected children from measles without causing the full-blown disease, the vaccine still wasn't weak enough.

Without weakening the virus further, the Boston team next tested the vaccine in developmentally disabled children at the Willowbrook State School, where Saul Krugman had performed his controversial experiments with hepatitis virus. On February 8, 1960, Katz injected twenty-three children with the vaccine, and twenty-three other children were given nothing. Six weeks later an outbreak of measles swept through Willowbrook, infecting hundreds of children and killing four. None of the vaccinated children got measles, but many unvaccinated children did. Enders's vaccine worked but, again, had caused a high rate of side effects. "[Representatives from] many pharmaceutical companies came and got material to produce vaccines," recalled Katz. "One was Maurice Hilleman."

HILLEMAN WAS EXCITED ABOUT WORKING WITH THE VACCINE DEVELoped by the Boston group. But he faced two difficult problems. Neither was easily solved.

The first involved side effects. Although the Enders team had passed David Edmonston's measles virus through several different human and animal cells, the virus still wasn't weak enough. Hilleman found, in experiments involving hundreds of healthy children, that half of those given Enders's vaccine had a rash, and most had fevers, some higher than 103 degrees. "It was toxic as hell," recalled Hilleman. "Some children had fevers so high that they had seizures. The Enders strain was the closest thing there was to a vaccine, but to me it was just an isolate. He hadn't made a vaccine." Even while he was worried about the safety of Enders's vaccine, Hilleman felt pressured by public health agencies anxious to prevent a disease that was killing thousands of American children every year. He had to find a way to make Enders's vaccine safer, and he had to do it quickly. The toxic-as-hell problem was solved by Joseph Stokes Jr., the pediatrician who had helped Hilleman test his mumps vaccine.

Hilleman chose Stokes because he was an expert on gamma globulin, the fraction of blood that contains antibodies. To make gamma globulin, Stokes took blood, let it clot, and stored it in the refrigerator. Eventually, red blood cells formed a clot at the bottom of the tube, and gamma globulin, contained in serum, floated to the top. Stokes knew that people infected with measles, mumps, polio, or hepatitis viruses rarely got the disease again. And he knew that antibodies were the reason why. In the mid-1930s, Stokes showed that gamma globulin taken from polio survivors protected children during polio epidemics. Ten years later, as a special consultant to the surgeon general during the Second World War, Stokes showed that gamma globulin from hepatitis survivors protected American soldiers from hepatitis. For his work on hepatitis, Stokes won the Presidential Medal of Freedom, the nation's highest civilian award.

Stokes proposed giving a tiny dose of gamma globulin along with Enders's measles vaccine, hoping to modify side effects. To see if the idea worked, Stokes and Hilleman went to a women's prison in central New Jersey.

Built in 1913 on a sprawling farm in rural Huntington County, Clinton Farms for Women was an ideal prison. An outgrowth of the turn-of-the-century prison reform movement, Clinton provided education, technical training, medical care, and a safe environment for inmates. In the early 1960s, Hilleman and Stokes made several visits to Clinton Farms. (During one early visit, they were eating lunch in the cafeteria when a waitress came to the table and asked them what they wanted to eat. Hilleman needed the prisoners to feel comfortable with his experiment, and he knew that the waitress was also a prisoner. "So what are you in for?" he asked, awkwardly, trying to make conversation. "I killed my parents," she replied. Seeing the stricken look on his face, she added, "But don't worry. You're safe here." Hilleman never asked that question again and, despite her assurances, never felt completely safe on subsequent visits.)

Edna Mahan, the prison's director, revolutionized life at Clinton Farms by taking the locks off doors and prohibiting guards from carrying guns. Prisoners could walk off the grounds any time they wanted. "The prisoners would leave the prison, travel down the road, get into a passing truck, and get themselves pregnant," recalled Hilleman. "The nursery was full of babies."

Stokes and Hilleman injected six infants with Enders's vaccine in one arm and gamma globulin in the other. None of the infants had a high fever, and only one had a mild rash. Encouraged, they tested hundreds of children. In the end, Stokes's gamma globulin strategy worked. Subsequent studies during the next few years showed that the percentage of children with rash decreased from 50 percent to 1, and with fever from 85 percent to 5.

Edna Mahan died in 1968, only a few years after Hilleman performed studies in her prison's nursery. In a small cemetery on the prison grounds, her elaborate tombstone is surrounded by forty tiny crosses, each representing babies who died in the prison from infections that are now easily prevented by vaccines.

THE SECOND PROBLEM THAT WORRIED HILLEMAN WAS THAT ENDERS'S vaccine might cause cancer. Although measles virus didn't cause cancer, Hilleman had reason to be concerned. His fears stemmed from an event that had occurred fifty years earlier.

In 1909 a farmer walked into the Rockefeller Institute in New York City carrying a dead chicken under his arm. Hands thickened by hard work, wearing thick bib overalls and heavy boots, the farmer watched as scientists, technicians, and graduate students milled through the lobby of one of the country's premier research institutes. Finally he got up the courage to ask where he could find Peyton Rous's laboratory. The farmer was certain that Rous, an expert in animal diseases, would know what had happened to his chicken.

A Baltimore-bred, Johns Hopkins–trained pathologist, Peyton Rous was thirty years old when he took the chicken from the farmer, laid it onto his laboratory bench, and dissected it. Just under the right breast was a large cancerous tumor. Rous found that the cancer had also spread to the liver, lungs, and heart. He asked the farmer whether any other chickens in his flock had a similar problem. "No," the farmer said, "only this one." Apparently, the cancer wasn't contagious.

Rous wanted to find what had caused the chicken's cancer. So he removed the tumor and carefully ground it up with sterile sand, completely destroying all the malignant tumor cells. Then he suspended the disrupted tumor cells in salt water and passed them through unglazed porcelain, which acted as a filter to trap bacteria. Rous found that when he injected the fluid that had passed through the filter into other chickens, tumors developed in those chickens as well. Within a few weeks, cancer had killed them all. "The [chickens] became emaciated, cold, and drowsy, and shortly died," he said. Because the tumors were caused by something passing through the filters, Rous knew that it couldn't be bacteria. And he knew that it couldn't be the cancer cells, because they had been destroyed by the sand and were too big to pass through the filters. It was something else. Rous reasoned that the agent that had passed through the filters was a virus.

On January 11, 1911, Peyton Rous, in a paper titled "Transmission of a Malignant Growth by Means of a Cell-Free Filtrate," was the first person to prove that viruses could cause cancer. Immediately other investigators tried to duplicate Rous's findings in mice and rats, but without success. Reasoning that cancer-causing viruses were at most a phenomenon unique to chickens, Rous gave up his investigations in 1915, and for the next four decades researchers relegated tumor viruses to the cabinet of freaks.

Although many cancer researchers working in the 1910s through the 1940s ignored Rous's findings, evidence continued to mount in favor of cancer-causing viruses. In the early 1930s Richard Shope, a veterinarian from Iowa, found that viruses caused giant warts on wild rabbits in the southwestern United States. (Although many consider them to be mythical creatures, jackalopes—jackrabbits with antelope horns—might be rabbits infected with Shope's wart-causing virus.) A few years later, researchers found a virus that caused tumors in the mammary glands of mice. But it wasn't until the 1950s, when Ludwik Gross, a Polish refugee, found a virus that caused leukemia in mice, that cancer-causing viruses came out of the cabinet and into the mainstream of virus research. Ten years later William Jarrett, working at the University of Glasgow in Scotland, found another virus that spread easily from one cat to another, causing leukemia. Not only did viruses cause cancer, but some cancer-causing viruses were also contagious.

In 1966, more than fifty years after he had examined the farmer's dead chicken, Peyton Rous won the Nobel Prize in medicine for "the discovery of tumor-inducing viruses." Rous received the prize, which cannot be awarded posthumously, when he was eighty-six years old.

. . . .

TODAY WE KNOW THAT SOME OF THE VIRUSES THAT CAUSE CANCER belong to the family of retroviruses, the most famous of which is human immunodeficiency virus (HIV), the virus that causes AIDS. When John Enders handed a vial of measles vaccine to Maurice Hilleman, it was loaded with a retrovirus that caused leukemia in chickens. Although they didn't know it, chicken leukemia virus had contaminated the eggs that the Enders team had used to make their vaccine.

At the time Hilleman wanted to make his measles vaccine, chicken leukemia virus—similar to the one that had infected Peyton Rous's chicken—infected about 20 percent of all chickens in the United States. The virus infected the liver, causing liver cancer; the kidneys, causing kidney cancer; ligaments, tendons, and skin, causing soft-tissue cancers; and cells of the immune system, causing leukemia and lymphoma. About 80 percent of chickens infected with chicken leukemia virus got leukemia. The virus was a tremendous headache for farmers, causing $200 million in lost revenue every year. Worse, in the early 1960s, researchers didn't know whether viruses that caused cancer in chickens could cause cancer in people. But they did know that animal retroviruses like chicken leukemia virus were capable of causing human cells in a test tube to become cancerous. "I wasn't going to license a vaccine with this virus in it," said Hilleman. "That would be the most unethical thing." Despite pressure from people such as Joe Smadel, director of the federal agency that licensed vaccines, to bring Enders's vaccine to market quickly, Hilleman refused to ignore the remote possibility that chicken leukemia virus could cause cancer in people. "Here was a vaccine that had been grown in cell culture, still highly virulent, and grown in cells that were loaded with leukemia virus," recalled Hilleman. "And I'd be damned if I was going to vaccinate all of these kids with [a virus causing] leukemia. I wouldn't do it. The government wanted to move forward because kids were dying [of measles]. When I told Smadel that we weren't going to bring out the [vaccine] with leukemia [virus] in it, he had a temper tantrum."

Although chicken leukemia virus was common, no one had found a way to detect it. And because contaminated eggs appeared perfectly normal, researchers couldn't tell which eggs were infected and which weren't. Hilleman was stuck. Fortunately, in 1961 a virologist at the University of California at Berkeley, Harry Rubin, found a way to detect chicken leukemia virus in the laboratory. "The Rubin test changed everything," recalled Hilleman. With Rubin's test now in hand, Hilleman could use eggs and chick embryos that didn't contain chicken leukemia virus to make his vaccine. First, Hilleman tried to breed his own flock of leukemia-virus–free chickens. But Merck was a company better suited to making drugs than to breeding chickens. So Hilleman turned to his friend Wendell Stanley for help. Stanley, who had won the Nobel Prize in medicine in 1946 for his work determining the structure of virus particles, directed Hilleman to a small farm in the Niles section of Fremont, California, Kimber Farms, where researchers had successfully bred a flock of leukemia-virus–free chickens. Hilleman found this hard to believe—he knew how much trouble he was having breeding them at Merck—but he was willing to give it a shot. So he got on a plane and flew to San Francisco. Then he drove forty miles to Fremont.

NILES WAS SMALL, CENTERED ON AN OLD FLOUR MILL, A SOUTHERN Pacific Railroad switching yard, some fruit packing plants, and a company that excavated sand and gravel.

Kimber Farms, part of the poultry-genetics mania that began in the early 1900s, was founded in the early 1930s by John Kimber, whose father was an Episcopal minister and mother a professional musician. Kimber was enormously successful, proving that the quality and size of an egg, the thickness of the shell, and the number of eggs produced could all be controlled by scientific breeding. He developed disease-free eggs, disease-resistant chickens, and hens that could lay two hundred and fifty eggs per year—accomplishments communicated to local farmers through his newsletter, *Kimberchik News*. But the notion of genetically breeding animals for food didn't sit well with critics, who saw Kimber's breeding operation as heartless and cruel. "Efficient, white-gowned workers in the antiseptic laboratories of Kimber Farms had little time for sentiment," recalled Page Smith and Charles Daniel, authors of *The Chicken Book*. "To them the baby chickens, half of whom were killed at birth and incinerated or fed to the hogs, hatched by the millions in their enormous incubators, [were] seen primarily as items on an assembly line. The fact that they were alive was, it seems fair to suggest, incidental."

To make chickens and eggs free of leukemia virus, Kimber scientists took eggs from hens that weren't infected with the virus, dipped them in organic iodine, and carefully put them into a sterile incubator. Male and female chickens born from these eggs made more chickens. Within a single generation, researchers at Kimber had bred chickens that were free of chicken leukemia virus. But it wasn't easy. The chickens were housed two hundred feet upwind from the nearest poultry house and screened to prevent contact with flies and rodents. Furthermore, caretakers had to put on protective clothing and shoes and step in a foot-pan containing disinfectant before entering the building. At the time, Hilleman didn't have the facilities or expertise at Merck to duplicate these procedures.

When Hilleman arrived at Kimber Farms, he walked into the main office and asked to speak to the principal investigator, Walter Hughes. He asked Hughes if he could buy some of his leukemia-free chickens. "That's our research flock," said Hughes. "I can't sell you those chickens." Hilleman considered his next move. "Do you have a boss?" he asked. Hughes escorted Hilleman into the office of the director of poultry research, W. F. Lamoreux. The result was the same. Lamoreux didn't want to sell his chickens. Hilleman tried harder: "One year from now there are going to be a lot of dead kids from measles, and you can do something to stop it." Lamereux wasn't moved. "We're not selling our chickens," he said. As Hilleman was leaving the office, he stopped, turned around, and tried one more time. Recognizing a familiar accent, he asked Lamoreux where he was from. "Helena," said Lamoreux. "Miles City," replied Hilleman, extending his hand. "Take them all," said Lamoreux, smiling broadly. "One buck apiece."

The first measles vaccine required a virologist and a chicken breeder. If both hadn't been born and raised in Montana, the road to a lifesaving vaccine might have been much longer.

HILLEMAN LATER SET UP HIS OWN FLOCK OF LEUKEMIA-VIRUS–FREE chickens on the Merck grounds, and between 1963 and 1968 he made millions of doses of Enders's measles vaccine. The vaccine worked. Despite the burden of its having to be given with gamma globulin, it decreased the incidence of measles in the United States. But Hilleman wasn't the only researcher, and Merck wasn't the only company to make measles vaccine. Two other pharmaceutical companies introduced their own vaccines. One, a vaccine made in dog kidneys by a veterinary vaccine maker, was on the market for only three weeks. "That vaccine was more dangerous than measles," recalled Hilleman. The other vaccine, made by killing natural measles virus with formaldehyde, was given to about a million American children before researchers found that immunity was dangerously

*For his work on the measles vaccine, Maurice Hilleman was interviewed in 1963 by Charles Collingwood of CBS. The segment was titled "The Taming of a Virus."*

short lived. Public health officials withdrew the killed measles vaccine after it had been sold for only four years.

Despite the success of Hilleman's measles vaccine, the need to give it with gamma globulin made it cumbersome to use. To solve the problem, Hilleman took Enders's measles vaccine and passed it forty more times through chick embryo cells. He called this new strain the Moraten strain, for *Mor*e *At*tenuated *En*ders. Merck first distributed the Moraten strain in 1968. Since then, it has been the only measles vaccine used in the United States. Between 1968 and 2021, hundreds of millions of doses have been given. As a result, the number of people infected every year in the United States has decreased from four million to fewer than fifty. Worldwide, the number of people killed by measles every year has decreased from eight million to about two hundred thousand." Measles vaccines save more than seven million lives a year. And the descendants of Kimber Farms's original flock of chickens, still maintained on the grounds of Merck, are used to make vaccines today.

What would have happened if Maurice Hilleman had made his measles vaccine using eggs contaminated with chicken leukemia virus? Would the vaccine have caused leukemia or other cancers? The answer came in 1972, ten years after Hilleman had licensed his first measles vaccine. Researchers studied about three thousand veterans of the Second World War who had died of cancer to see whether they were likely to have received a yellow fever vaccine made in eggs contaminated with chicken leukemia virus. The answer was no. Although chicken leukemia virus caused cancer in chickens, it didn't cause cancer in humans. But when Maurice Hilleman made his measles vaccines, he didn't know that. "I just couldn't take that chance," he recalled.

# CHAPTER 5

# Coughs, Colds, Cancers, and Chickens

*"This blender is going to revolutionize American drinks."*

Fred Waring

Between 1944, when he left the University of Chicago to work at E. R. Squibb, and 1968, when he made his own measles vaccine at Merck, Maurice Hilleman made or tried to make several unusual vaccines.

One prevented cancer.

Hilleman's vaccines to prevent measles, mumps, and pandemic influenza all had one thing in common: they were made in chickens. "Chickens were my best friend," he said. "I could hypnotize them. All you had to do was lay them on their side and let them stare at a white feather. They were transfixed by that feather. They helped me so much. Maybe I could do something for them." When Hilleman finally paid his debt to chickens, he also fulfilled a promise that was made, but never kept, by Herbert Hoover.

During the 1928 campaign for the presidency of the United States, newspaper advertisements by the Republican National Committee proclaimed that Warren Harding and Calvin Coolidge had "reduced [work] hours and increased earning capacity, silenced discontent, put the proverbial 'chicken in every pot,' and a car in every backyard, to boot." The advertisement stressed that a vote for Herbert Hoover was a vote for continued prosperity. The promise of a chicken in every pot was attractive. Chickens, like turkeys today, were expensive, served only as a delicacy for special occasions. But Herbert Hoover, who presided over the stock market crash of 1929 and the beginning of the Great Depression of the 1930s, never kept the promise.

Hilleman remembered a strange disease that had affected chickens on his family's farm when he was a child: "Every year we had chickens that would die or become ill from an unknown cause." Some chickens destined for the Hillemans' dinner table were thin and weak or had ghastly, rock-hard tumors under their skin or in their organs. "[Aunt Edith] would have a chicken killed now and then. When she saw a chicken that had lumps on the skin or had any tumors, she would say 'We can't eat that.'" Years later, the mysterious ailment had a name: Marek's disease. By the early 1960s, researchers found that Marek's disease was caused by a herpesvirus.

Marek's disease wasn't a problem only on the Hilleman farm. It was a problem on many farms, affecting 20 percent of all chickens produced in the United States. The disease attacked the nerves of the legs, causing paralysis; chickens died because they couldn't get food and water or because they were trampled by other birds. Farmers called it "range paralysis," referring to infected chickens as being "down in the leg." The virus also caused cancer of the skin, ovaries, liver, kidneys, heart, and spleen, killing one third of its victims. There was no treatment. Farmers simply culled infected chickens from the flock and destroyed them.

Marek's disease was also highly contagious. The virus hid in the fine, light dandruff that filled the air of the chicken coop. "We had these chickens dying all the time with range paralysis," recalled Hilleman. "[In the chicken coop] there were these great festoons of chicken dandruff. We had a wire hanging down from the ceiling, and it would be just like a hive of bees swarming." Because of its surface charge, the dandruff would cling to the wires. "Electrostatically, you would have one or two gallons of chicken dandruff hanging on there." The light spray of dandruff hung in the air for months, easily spreading from coop to coop and farm to farm.

Hilleman had seen Marek's disease and remembered it. When Ben Burmester, a veterinary researcher in Michigan, found that a herpesvirus similar to Marek's virus also caused disease in turkeys and quail, Hilleman saw his opening. "One day I get a call from Burmester from the East Lansing Regional Poultry Center," recalled Hilleman. "[He] said, 'Maurice, we've isolated a virus here from turkeys, and when we vaccinate chickens, they become resistant to Marek's disease.' I said 'Ben, I'll be out tomorrow.' I asked him, 'What is your interest?' And [Ben] told me, 'I just made an observation, I can't do anything with it.'"

Hilleman took Burmester's turkey herpesvirus, grew it in the laboratory, injected it into one-day-old chicks, and found that they were protected from Marek's disease. But before he could distribute his Marek's vaccine, Hilleman had one more obstacle. He had to convince Merck's board of directors to get into the chicken vaccine business. Max Tishler, Hilleman's boss and the director of research at Merck, set up a meeting with the board. Tishler wasn't interested in making products for animals and was certain that the board would agree. "I got hell [from Tishler] for developing the product," recalls Hilleman. "I was asked to go to the board of directors, and the directors said, 'That's wonderful; go ahead.' Max came running after me after the meeting and said, 'Why did you do that?' I said 'You were the one who told me to go to the meeting.' And [Tishler] said, 'Yeah, but I didn't want you to succeed. We're not in the chicken business.' [Marek's disease] vaccine was such a drain on our laboratory. But it was the world's first cancer vaccine." Merck, like it or not, was now selling products made for chickens. Soon it would be selling the chickens too.

Chicken farmers are in the business of converting carbohydrate (grain) into protein (meat). "There are two kinds of chicken breeders in this world," recalled Hilleman. "There are those who breed chickens to get maximum egg production, and those who breed for maximum meat production. They all, with the exception of one company, would breed for resistance to Marek's disease." The one company was Hubbard Farms of Walpole, New Hampshire.

In 1921 Oliver Hubbard, one of the first students to graduate from the University of New Hampshire with a degree in poultry farming, founded Hubbard Farms. By the early 1930s, Hubbard had developed the New Hampshire chicken, a breed unrivaled for egg and meat production. But there was one problem. The New Hampshire chicken was more susceptible to Marek's disease than any other breed. "Hubbard Farm chickens were by far the most productive means of converting carbohydrate to protein than any other chicken in the world," remembered Hilleman. "But if you got Marek's into the flock, you were done." Hilleman saw an opportunity for Merck. "Merck was out for an acquisition, and it was obvious what to do. We'll buy Hubbard! [We'll combine] these efficient carbohydrate-to-protein chickens with the vaccine that compensated for their genetic susceptibility to Marek's disease."

In 1974, Merck bought Hubbard Farms for $70 million. Hilleman's vaccine, the first to prevent cancer in any species, revolutionized the poultry business. Excess production caused the price of chickens to drop from $2 per broiler to forty cents and of eggs from fifty cents per dozen to five cents. Soon everyone could afford chickens, and for a while, Merck, one of the more conservative pharmaceutical companies in the United States, was the biggest chicken and egg producer in the world.

WITH HIS MAREK'S VACCINE, MAURICE HILLEMAN BECAME THE FIRST person to make a vaccine to prevent cancer. He was also the first person to purify, characterize, and produce a drug that is now used to treat certain cancers in people.

In the early 1900s Alexander Fleming, a Scottish biologist working in London, returned from a vacation to find something unusual in his laboratory. Fleming worked with *Staphylococcus aureus*, a bacterium commonly found on the skin and in the environment. He grew the bacteria in laboratory plates, where it formed small golden-yellow colonies (in Latin *aureus* means "gold"). When Fleming returned to the laboratory after several weeks he found, much to his dismay, that the plates had been overrun with mold: fluffy green mold. But he also noticed something else. Although many colonies of bacteria were present throughout the dish, he didn't see any in the areas immediately surrounding the mold. Fleming reasoned that a substance produced by the mold was killing the bacteria. Because the name of the mold was *Penicillium notatum*, Fleming called the substance penicillin. In 1929 Fleming published a paper titled "On the Antimicrobial Action of Cultures of a Penicillium with Special Reference to Their Use in the Isolation of B. Influenzae." Fleming's first description of penicillin is regarded as one of the most important medical papers ever written.

For the next six years Alexander Fleming worked fitfully on his new antibiotic. But Fleming was a biologist, not a chemist; he never successfully purified penicillin. In 1935, six years after publishing his discovery, he gave up. Several years later, at the start of the Second World War, a team of researchers at Oxford University headed by Howard Florey picked up where Fleming had left off. They purified penicillin, described its physical and chemical properties, studied its effects in animals and humans, and showed how to mass-produce it, just in time to save the lives of tens of thousands of Allied soldiers.

Ten years after abandoning his research on penicillin, Alexander Fleming won the Nobel Prize in medicine "for the discovery of penicillin and its curative effect in various infectious diseases." Although our understanding of what penicillin is, how it works, and how it can be used to save lives couldn't have happened without Howard Florey, few know his name. When people think of penicillin, they think of Alexander Fleming. The story of Fleming and Florey would be repeated with the discovery of the first substance to inhibit the growth of viruses and treat cancer.

In 1957 Alick Isaacs, a Scottish virologist working on influenza virus, teamed with a Swiss biologist named Jean Lindenmann. Working in Mill Hill Laboratories just outside of London, they found, as had many researchers before them, that influenza virus destroyed cells in the membrane that surrounded chick embryos. But unlike other researchers, they found that if they first treated the chick cells with a strain of killed influenza virus, then live influenza virus didn't

*Maurice Hilleman and the interferon team at Merck, circa early 1960s.*

destroy the cells. Isaacs and Lindenmann reasoned that chick cells exposed to killed influenza virus were making a substance that inhibited the ability of influenza virus to grow. They called this interfering substance *interferon*. Isaacs thought that biologists had now described their own unique element. "It is time that biologists had a fundamental particle," he said, "for the physicists have so many: electron, neutron, proton." Many skeptical scientists disagreed, preferring to call Isaacs and Lindenmann's finding an "imaginon."

Unfortunately, Isaacs and Lindenmann couldn't purify interferon, so they couldn't study what it was or how it worked. Hilleman, by figuring out how to mass-produce it, revolutionized the field of interferon research. Where Isaacs and Lindenmann's preparation contained seventy units of interferon per milliliter (one fifth of a teaspoon of liquid), Hilleman's contained more than two hundred thousand. Hilleman's final product was so pure that one unit of interferon activity was contained in only forty nanograms of protein (about one two-millionths of the weight of a grain of sand)—a potency to treat viruses that was, on a weight basis, actually greater than the potency of penicillin to treat bacteria.

Because he was the first to purify interferon, Hilleman was the first to detail its physical, chemical, and biological properties. He found that interferon was produced not only by cells from chick embryos but also by cells from calves, hamsters, dogs, rabbits, mice, and humans. He found that interferon inhibited the growth of many human and animal viruses, including cowpox, rabies, and yellow fever. He found that interferon not only prevented infections caused by viruses but also prevented cancers caused by viruses. "Interferon was the first antiviral agent, the grandfather of them all. It was a real good inducer of resistance," recalled Hilleman. "We could stop [cancer-causing] viruses. We could stop nearly every god-damned virus."

In the mid-1960s, Maurice Hilleman reasoned that interferon could be useful in treating chronic infections and cancers. And he was right. Today interferon is used in the treatment of chronic infections with hepatitis B and hepatitis C viruses, as well as cancers such as leukemia, lymphoma, and malignant melanoma.

Hilleman made one of his vaccines in a leaky Waring blender.

In 1944, when the United States was preparing to invade the Far East, the military became very concerned about a virus called Japanese encephalitis virus (JEV), one of the most common infections of the brain in the world. Transmitted by mosquitoes, JEV causes seizures, paralysis, coma, and death in one of every three of its victims; another third are left with permanent brain damage. Indeed, JEV is still a common infection in Southeast Asia. Every year the virus infects twenty thousand people—mostly children—and kills six thousand.

Because American servicemen had never been exposed to JEV, they were, like the children of Asia, highly susceptible to the disease and its consequences. Military health officials asked pharmaceutical companies to submit bids for the production of JEV vaccine. Hilleman had just taken a job at E. R. Squibb. He wanted Squibb to win the contract. While at the University of Chicago, Hilleman had found that JEV could be grown in mouse brains and killed with formaldehyde. He also knew from studies performed in Russia and

Japan that formaldehyde-treated JEV could prevent disease. "We [at Squibb] put out a bid for $3 a dose and said that we could have a production facility up and running in thirty days," he said.

Hilleman promised to make hundreds of thousands of doses of JEV vaccine for the military. But Squibb didn't share his patriotic enthusiasm, certain that Hilleman couldn't possibly make that much vaccine that quickly. "We had nothing on the plant site other than an old horse barn," recalled Hilleman. "And we had to start production in thirty days. They gave me an engineer. He said to me, 'How in the hell are we going to start production in thirty days?' So we sat down and tried to figure out how to make a laboratory. We bulldozed all of the horse manure out of the barn and painted a concrete floor. We cleaned out the loft upstairs, where all the hay had been stored and put in a stairway. Then we put in heating and electricity."

To make the vaccine, Hilleman's technicians injected JEV into mouse brains. Then they waited a few days until the virus reproduced. Hilleman remembered what happened next: "The girls would take the mice and [kill] them with ether. Then they would dip the mice in Lysol, strip the skin from the skull, and scoop out the brains with scissors." Before treating the virus with formaldehyde, Hilleman put hundreds of tiny mouse brains into a Waring blender (named for the popular bandleader Fred Waring, who marketed it). Sometimes, when homogenized brains leaked out of the blender, Hilleman worried that technicians would catch the disease. "We put the brains into Fred Waring's cocktail blender. And those damned blenders would leak out of the base, and sometimes brains would squirt out of the tops of the blenders. Waring didn't give a damn about cocktail loss, but when mouse brains started to leak onto the floor, it scared the hell out of me." (For whatever reason, television commercials for the Waring blender in the 1950s never featured its capacity to homogenize mouse brains.) Thirty women, working eight-hour shifts and processing two mice per minute, harvested about thirty thousand mouse brains a day. Because the JEV vaccine was given as a series of three doses, it took about three months to make enough of it to immunize six hundred thousand

American troops. Military epidemiologists never performed studies to determine whether Hilleman's JEV vaccine worked during the latter stages of the Second World War. But it is likely that the vaccine prevented disease in thousands of soldiers. JEV vaccine production in mouse brains was discontinued in 2006.

ALTHOUGH HILLEMAN HAD A REMARKABLE RECORD OF SUCCESS IN making vaccines, he failed to make one to prevent the world's most widespread and most annoying infection: the common cold. But he tried.

The common cold has plagued humankind since the beginning of recorded history.

Countless efforts to cure it have generally failed. In the fifth century B.C., Hippocrates noted that despite many attempts, therapeutic bleeding didn't work. In the first century A.D., Pliny the Elder recommended "kissing the hairy muzzle of a mouse" (which also didn't work). In the eighteenth century, Benjamin Franklin said that colds were spread from one person to another (which was true) and that people would be less likely to catch colds if they avoided damp, cold conditions (which wasn't true). Today, people treat colds with echinacea, St. John's wort, vitamin C, vitamin E, and zinc. But perhaps the best advice came in the nineteenth century from the renowned physician William Osler. "There is just one way to treat a cold," he said, "and that is with contempt."

Colds account for half of all acute medical conditions. But despite tremendous technological advances in isolating, identifying, sequencing, and cloning cold viruses, as well as advances in understanding how the immune system responds to these viruses, scientists and researchers have done nothing to prevent the common cold.

Most of our understanding of the common cold has been an indirect consequence of the Second World War. When the U. S. Army evacuated Britain at the end of the war, it left behind an army hospital in Salisbury, England, that had been staffed by doctors from Harvard Medical School. Christopher Andrewes, a British researcher who was the first to isolate influenza virus from people, set up the

Common Cold Research Unit in the abandoned buildings. During the next four years, Andrewes persuaded two thousand adults to take "ten-day holidays" at his research unit. He wanted to see what would happen if he inoculated volunteers with nose and throat washings taken from people suffering from colds. (On its face, this doesn't sound like much of a holiday.) Andrewes found a few things that surprised him. About half of those inoculated with cold viruses came down with colds; women were more likely to catch colds than men; antihistamines, which had just been developed, were worthless; and people who were cold were not more likely to catch colds—they were actually less likely. Andrewes figured out the last point by asking people who had been inoculated with cold viruses to stand in a draft for thirty minutes in a wet bathing suit. He also found that people inoculated with a cold virus from one person weren't protected months later when challenged with a virus from someone else.

The first apparent breakthrough in preventing the common cold came in the fall of 1953, when Winston Price, a thirty-four-year-old biochemist working at Johns Hopkins Hospital, isolated a virus from the nose of a student nurse. He called it JH virus, for Johns Hopkins. Then he took the virus, grew it in monkey kidney cells, killed it with formaldehyde, and injected it into the arms of a hundred boys in a local training school. The results were dramatic. During the next two years, children who had received Price's vaccine were eight times less likely to get a cold than those who hadn't. Price was cautious. "It's absolutely misleading if anyone thinks we are going to have an all-inclusive cure for colds," he said. "This is just the opening wedge, the first piece out of the pie, and an opening we have not had before. What we hope is that by using similar methods we may help isolate one or more viruses which make up the other part of the cold pie."

Scientists and physicians hailed Price's findings as groundbreaking. George Hirst, director of the Public Health Research Institute in New York City, said that "the work by Dr. Price on the new JH virus is a promising lead in the attack on the common cold." By the end of the year, one vaccine maker said that it would soon have a vaccine to prevent colds. But to make a vaccine, researchers needed to find out how many different viruses caused the common cold. In the early 1960s, Maurice Hilleman answered the question. Hilleman swabbed the throats of Merck employees, students at the University of Pennsylvania, and children admitted to the Children's Hospital of Philadelphia. He also collected samples of cold viruses from other researchers. Then he grew the viruses in laboratory cells, obtained blood from people who had just been infected, and tested the serum to see whether the viruses were immunologically similar or distinct. Hilleman identified fifty-four different types of cold viruses; forty-one of these were first isolated in his laboratory. Further, Hilleman found that natural infection with one type of cold virus protected against disease caused by that same virus for at least four years, but it didn't protect against disease caused by other types. The common cold was common, not because immunity was short-lived but because there were so many different cold viruses. "If there was just one type, then a vaccine could protect against colds for the rest of your life: just like the measles and mumps vaccines," said Hilleman.

Hilleman tried to make a common cold vaccine by putting different types of cold viruses into a capsule. On May 26, 1965, in concert with the ethic of the time, he fed his vaccine to nineteen developmentally disabled children in the Vineland State School in New Jersey. Hilleman obtained frequent X-ray films of their stomachs to determine when the capsules opened. Although swallowing live cold viruses didn't cause any symptoms, no one developed antibodies; Hilleman's attempt to make a common cold vaccine had failed. Then Hilleman tried to find some degree of similarity among the different cold viruses that he could exploit to make a vaccine. But he never found it. "There was just no crossing at all between these strains," he said. Today, more than a hundred different types of cold viruses have been found, and no one has been able to make a vaccine to prevent them.

So why did the vaccine trial conducted in Baltimore by Winston Price work? If there are at least a hundred different types of cold viruses, why did a vaccine that contained only one strain of virus cause such a dramatic decrease in the incidence of colds for two years? The truth is that it didn't. "His study was a complete fraud," said Hilleman. "He made up his data. I found out about it when I was working at Walter Reed." No one has ever duplicated the success found by Winston Price in the mid–1950s, and given the unlikelihood of using one type of virus to protect against a disease caused by more than a hundred different strains, no one probably ever will.

Although he failed to make a vaccine to prevent the common cold, Hilleman had shown why it was so common. Next he set his sights on a virus that most doctors ignored, unless it infected pregnant women.

# CHAPTER 6

## The Monster Maker

*"If you cannot have what you believe in you must believe in what you have."*

George Bernard Shaw

In the spring of 1941 two mothers struck up a conversation in the waiting room of a doctor's office in Sydney, Australia. Both were holding babies on their laps. The mothers soon discovered that they were there for the same reason—their babies were blind. Looking for a clue to their misfortune, they compared pregnancies. Neither had traveled outside of Sydney, neither had relatives with eye problems, both had eaten well, and both had taken their vitamins faithfully. They had one more thing in common: both had been infected with German measles early in their pregnancies. Sorting through papers at the nurse's desk, the ophthalmologist, Norman McAlister Gregg, overheard the conversation. Gregg thought of German measles as a trivial infection of childhood. He couldn't believe that it caused blindness.

First described by German doctors as a disease similar to measles, German measles was later named *rubella* by a British physician during an outbreak in an all-boys boarding school in India. The doctor reported that students first experienced an uncomfortable swelling of the lymph glands behind their ears and on the backs of their necks. In a few days they were weary with fever and pinkeye. Later, a rash—red and slightly raised from the skin—appeared at the hairline and spread to the rest of the face. (In Latin, *rubella* means "little red.") The rash was barely noticeable, the fever was slight, and the fatigue was so mild that only a few boys missed school. Compared with measles, chickenpox, and scarlet fever—other diseases that caused rash and fever—rubella seemed to be the mildest of all childhood infections.

For the next few weeks Gregg examined the medical records of every woman in his care who had a baby with birth defects. He knew that two years earlier, in 1939, a rubella epidemic had swept across the continent. As he sorted through the records, Gregg had the uncomfortable sense that beginning nine months after the epidemic, he was caring for more and more babies who were blind. Was it possible that rubella had damaged babies in their mothers' wombs? Gregg found seventy-eight mothers whose babies were blind; sixty-eight had had symptoms of rubella early in their pregnancies.

In 1941, in a now landmark paper titled "Congenital Cataracts Following German Measles in the Mother," Norman McAlister Gregg published his findings in a little-known, little-read medical journal, *Transactions of the Ophthalmological Society of Australia.* He was fifty years old. Gregg had never published a scientific paper, was unknown to medical researchers, and lived on a continent far away from influential medical centers in the United States and Europe. His anonymity, coupled with the fact that he was proposing something that had never been proposed before—that a virus could cause birth defects—cast doubt on Gregg's observations. Few believed him.

Although many researchers were skeptical, some were intrigued by Gregg's hypothesis. In the twenty years following his observation, researchers in Australia, Sweden, England, and the

United States confirmed and extended his findings. They found that not only did rubella infection during pregnancy cause blindness but it also caused heart defects and deafness. (Viruses or drugs that damage babies in the womb are called *teratogens*—literally "monster makers.")

Although epidemics had occurred throughout the twentieth century, Americans didn't experience the full horror of rubella until the early 1960s. Between 1963 and 1964, in one of the worst epidemics ever recorded, rubella infected twelve million Americans. Among those infected were thousands of pregnant women. Rubella virus killed six thousand fetuses soon after conception and two thousand more at birth. The virus permanently harmed another twenty thousand unborn babies by infecting the liver, causing hepatitis; the pancreas, causing diabetes; the lungs, causing pneumonia; and the brain, causing mental retardation, deafness, blindness, epilepsy, and autism. Knowing that eight of every ten pregnant women infected with rubella early in pregnancy would give birth to babies who were severely harmed by the virus, mothers were left with a Sophie's choice: deciding whether their unborn babies should live or die. Unwilling to take odds that were so heavily stacked against them, five thousand women chose to abort their pregnancies. Many would never conceive again.

While rubella virus was sweeping across the world in the early 1960s, the war in Vietnam had just begun. When it was over, ten years later, fifty-eight thousand Americans had lost their lives. At home, rubella virus killed or wounded thirty thousand children in one year. But unlike the Vietnam War, the war waged by rubella against American children was unaccompanied by news bulletins, demonstrations, or lively debates in Congress. And not a single shot was fired.

Although rubella virus was the first infection found to cause birth defects, it wasn't the last. Bacteria such as *Treponema pallidum,* which causes syphilis; parasites such as *Toxoplasma;* and other viruses such as the chickenpox virus all cause birth defects. But no organism is more common, more thorough, or more consistent in its destruction of unborn children than rubella virus.

. . . .

In the early 1960s, a few years before the worst rubella epidemic in history, Maurice Hilleman began to work on a rubella vaccine. Hilleman knew that massive outbreaks of rubella had occurred in the United States in 1935, 1943, 1952, and 1958—about one every seven years. While working on his vaccine he witnessed the rubella epidemic of 1963–1964. He anticipated that the next one would arrive between 1970 and 1973. To have any hope of saving the lives of unborn children, he would have to make a rubella vaccine quickly.

Hilleman started by capturing rubella virus from the throat of an eight-year-old Philadelphia boy whose last name was Benoit. The vaccine would be known as the Benoit strain. He weakened the virus by growing it in monkey kidneys and duck embryos. On January 26, 1965, Hilleman injected his rubella vaccine into the arms of developmentally disabled children in group homes in and around Philadelphia. All developed rubella antibodies, and none had symptoms of infection. Months later, when a small epidemic of rubella swept through Pennsylvania, Hilleman found that 88 percent of unimmunized children got rubella, but all who had received his vaccine were protected. Confident, he couldn't wait to test it in more children. But Hilleman's efforts would soon be thwarted by someone he had heard of but never met—someone who, although not a researcher, doctor, politician, pharmaceutical company executive, or public health official, was more powerful than any one else in science or medicine.

Her name was Mary Lasker, the wife of advertising executive Albert Lasker. Albert "had a new gimmick to make money, and that was to work for nothing," recalled Hilleman. Immediately after graduating from high school, Albert joined the advertising agency of Lord and Thomas in New York City. Figuring that no one would be impressed by his youth, he offered to work for free, asking only to be paid in his clients' stock. When some clients became Fortune 500

*Mary Lasker stands between President John F. Kennedy and Vice President Lyndon B. Johnson at the White House, April 11, 1961, during a meeting of the Committee on Equal Employment Opportunity (courtesy of the Bettmann Archives).*

companies, Albert Lasker became a multimillionaire. At twenty-eight, he owned Lord and Thomas; two years later, he retired. In 1940, when he was sixty years old and she was forty, Albert married Mary Woodard. When they met, Mary was working at the Reinhardt Galleries in New York City, setting up private loan exhibitions of French master painters. A native of Watertown, Wisconsin, Mary had graduated from Radcliffe in 1923 and studied briefly at Oxford before returning to the United States. In 1942, Mary convinced Albert to create the Mary and Albert Lasker Foundation.

Although the twentieth century witnessed many philanthropists interested in social change, few had the indomitable will, courage, or resources of Mary Lasker. Her generous gifts to Planned Parenthood in the 1930s and 1940s made her the principal source of funding for the birth control movement in the United States. But her real passion was medical research. Lasker was responsible for a federal bill in 1971 that made the conquest of cancer a national goal. And she was a dominant force behind the creation of the National Cancer Institute, the first institute within the National Institutes of Health. Scientists and the media adored Mary Lasker. Jonas Salk said, "When I think of Mary Lasker, I think of a matchmaker between science and society." Michael DeBakey, the inventor of artificial hearts and pioneer in heart transplants and Mobile Army Surgical Hospitals (MASH units), said, "The National Institutes of Health has flowered because in many ways [Mary Lasker] gave birth to it and nursed it. It was in existence, but it was she who got the funding for it." *BusinessWeek* called her "the fairy godmother of medical research." Lasker won the French Legion of Honor, the Presidential Medal of Freedom, and the Congressional Gold Medal. She also established the Lasker Award, the most prestigious prize for biomedical research in the United States. (Its winners often later win the Nobel Prize.) But despite all her good works, Maurice Hilleman feared Mary Lasker. "You have to credit Mary Lasker for doing all of that," he said. "But she could kill you."

Lasker called Merck and asked Max Tishler, Hilleman's boss and the president of Merck Research Laboratories, to come to New York for a meeting. Lasker wanted Tishler and Merck to stop working on their rubella vaccine. She knew that Harry Meyer and Paul Parkman, both of whom worked at the Division of Biologics Standards (the agency responsible for licensing new vaccines in the United States) were making their own vaccine by taking rubella virus from an army recruit and passing it seventy-seven times through monkey kidney cells. Lasker reasoned that because Meyer and Parkman worked for the licensing agency, their vaccine would be licensed more quickly than Hilleman's, and she didn't want competition to slow the process. Initially, Lasker wanted to meet with Tishler alone, but Tishler refused, saying that he would meet Lasker only if Hilleman were there too. "I got a call from Max Tishler one day," said Hilleman, "and he said, 'Do you know this guy Harry Meyer?' Mary Lasker says that he has developed a rubella vaccine. Mary has talked to some people, and she thinks it could be a pretty good vaccine. [She] wants us to come up to her apartment in New York City and tell us what she thinks needs to be done." In Tishler, Hilleman now had a powerful ally.

. . . .

Like Mary Lasker, Max Tishler was used to getting what he wanted. Born in 1906 to European Jewish immigrants, the fifth of six children, Tishler survived a difficult childhood. His father, a cobbler, abandoned the family when Max was only five years old. To supplement the family's income, Max went to work by selling newspapers, working as a pharmacist's assistant, and delivering doughnuts for a local bakery. He was an outstanding student. Winning several awards and scholarships, Tishler later graduated from Tufts College and Harvard University, where he got a doctorate in chemistry. But positions in academia were scarce. So in 1937 Tishler landed a job at a growing pharmaceutical company in Rahway, New Jersey: Merck. He was attracted to Merck because of its solid revenues from chemicals such as iodine, silver nitrate, ether, and chloroform; because it was one of the few chemical companies in America that would hire a Jew; and because its president, George Merck, was interested in innovation.

Tishler's genius, combined with Merck's resources, led to an unparalleled series of successes. Tishler figured out how to mass-produce riboflavin (vitamin $B_2$) and pyridoxine (vitamin $B_6$), allowing food makers to add vitamins to enrich white bread. In 1942, after doctors found that penicillin treated deadly infections, Max Tishler was one of the first Americans to make it. In 1948, when a young researcher at the Mayo Clinic found a hormone, cortisone, that treated painful, achy joints, Tishler found a way to synthesize large quantities. In addition to penicillin, Tishler made other antibiotics. One, sulfaquinoxaline, treated a bacterial infection of chickens; when the drug was added to their feed, more chickens made it to market and sold for less.

Short, with red curly hair, horn-rimmed glasses, and a raspy voice, Max Tishler was a difficult man. "Max was driven to do things well," recalled a co-worker, "and he couldn't tolerate problems not being solved. He was utterly fearless in the face of trouble and actually impatient to hear all the bad news—all of the failure of good ideas, or setbacks from whatever source. Unlike most of us, who seem to need a little time to face up to reversals, he never even blinked." Roy Vagelos, former chief executive officer at Merck and an early pioneer of cholesterol-lowering drugs, also remembered Max Tishler. "So the story goes, and it's a true one; Max had the group geared up to isolate vitamin $B_{12}$, which is ruby red. It was isolated from tons of livers, and it was a breakthrough at Merck. People were working around the clock to get it done. Max had the habit of wandering around the laboratory at any time and would just burst in to ask you questions. So, one time in the middle of the night, these guys, having purified the substance, were putting it through a pressure filtration, and they were squeezing it through this tube. The tube broke and it started to leak. The door opened and Max walked in, and he looked at the red stuff on the floor and looked at them. They were sweating profusely. 'I hope that's someone's blood,' he said."

IN THE SPRING OF 1966, HILLEMAN AND TISHLER TOOK A TRAIN FROM Philadelphia to New York City and a taxi to Lasker's newly built apartment overlooking Central Park West. The apartment house was a fashionable address for the powerful elite, including novelist Truman Capote, Attorney General William Rogers, and Senator Robert F. Kennedy. Lasker ushered Hilleman and Tishler into rooms decorated with paintings by Miro, Renoir, Cezanne, and Dali. The scientists sat at the dining room table, quietly and nervously, waiting to hear what Lasker had to say.

Lasker explained that she had been haunted by the recent rubella outbreak. She said she was proud that Merck had chosen to work on a vaccine, but she also knew that Harry Meyer had a vaccine, and she feared that competition would only delay development. Hilleman slowly realized what Lasker was getting at: she was going to ask him to abandon his vaccine. "I explained that we needed to make vaccine to avoid the next epidemic," recalled Hilleman. "She said, 'I don't think we're going to be able to do it in time because you've got two vaccines competing. And one of these vaccines is made by the

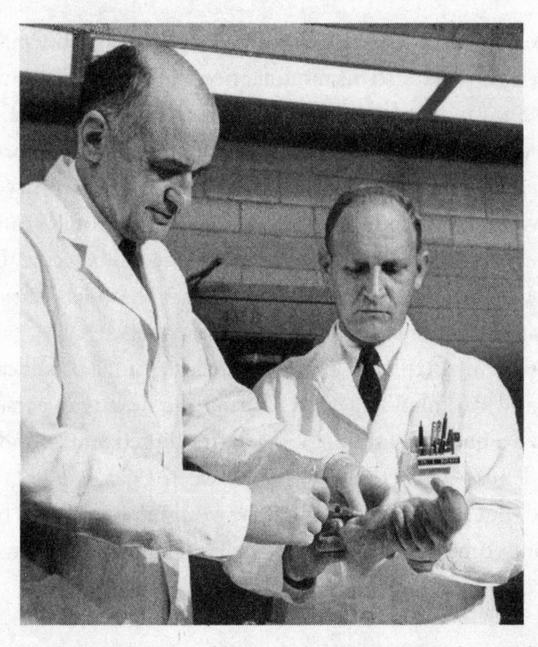

*Maurice Hilleman, assisted by coworker Eugene Buynak, inoculates a duck with rubella virus, circa late 1960s.*

federal regulator. Just whose vaccine do you think is going to get approved first?'" Hilleman remembered thinking, "Meyer's vaccine is [not] going to be approved first because he didn't make it into a vaccine yet. It was just a goddamned experiment." Lasker asked Hilleman and Tishler to go back to Merck and think about her request.

Standing on the street in front of Lasker's apartment after the meeting, Tishler turned to Hilleman. "I'll do whatever you want," said Tishler. "You tell me." Hilleman, for one of the few times in his life, succumbed to the pressure. "I'll tell you what I will do," said Hilleman. "I'll get the [vaccine] virus from Meyer and see what I can do." Hilleman took Meyer's virus and injected it into children living in and around Philadelphia. But he found that the virus had side effects that were intolerable. "I got [Meyer's vaccine] and put it into about twenty kids, and, Jesus Christ, it was awful: toxic, toxic, toxic. So I passed it five more times in duck [embryos] and attenuated it."

One year after the meeting with Mary Lasker, Hilleman compared his own vaccine to his modification of Meyer's. He found that both vaccines induced antibodies and that both were safe. But he also found that his vaccine induced much higher levels of protective antibodies. Hilleman had a choice. He could either proceed with his own vaccine, which he knew was better, or he could grant Mary Lasker her wish. In 1967, Hilleman abandoned his vaccine. Later he said, "If I have only one regret in my life, it's that I let Mary Lasker talk me out of making my own rubella vaccine."

By 1969, Hilleman had obtained a license for his modification of Meyer's and Parkman's vaccine. During the next ten years, Merck distributed a hundred million doses in the United States, and the rubella epidemic—expected to occur between 1970 and 1973—never happened. But the first rubella vaccine wouldn't be the last, because, for the only time in his life, Maurice Hilleman's vaccine would be replaced by a better one.

WHILE MAURICE HILLEMAN AND HARRY MEYER WERE TRYING TO make their rubella vaccines, Stanley Plotkin was making his. Meyer had used cells from monkeys, and Hilleman had used cells from ducks. Plotkin used cells from an aborted human fetus. His choice opened a door to a controversy that has never closed.

Stanley Plotkin is an energetic scientist with an avuncular style and disarming wit. Born in the Bronx, New York, he attended the Bronx High School of Science, a highly competitive school for gifted teenagers. "I've never had as much intellectual competition as I did in that school," recalled Plotkin. "Not in college, medical school, or research." Plotkin was inspired to a life of science by Sinclair Lewis's novel *Arrowsmith*, the story of Martin Arrowsmith, a boy from Minnesota who becomes a family doctor in the Midwest. Intellectually dissatisfied with his work, Arrowsmith is drawn to the fictional McGurk Institute in New York City and the wise, even-handed tutelage of a German Jewish scientist named Max Gottlieb. With Gottlieb's help, Arrowsmith discovers a virus that kills bacteria—a finding that he reasons will revolutionize the treatment of bacterial infections. (Sinclair Lewis wrote *Arrowsmith* ten years before the discovery of antibiotics.) Stanley Plotkin's life would soon parallel Martin Arrowsmith's.

After graduating from high school with honors, Plotkin attended New York University and then Downstate Medical Center in Brooklyn, both on full scholarships. Later he joined the Epidemic Intelligence Service (EIS) at the Communicable Diseases Center in Atlanta, now the Centers for Disease Control and Prevention. Alexander Langmuir, the head of the EIS, wanted Plotkin to study anthrax, a fatal respiratory disease, so he sent him to the city with the greatest incidence of anthrax in the United States: Philadelphia. "Anthrax was a problem in Philadelphia [because] the clothing industry [for which Philadelphia was an important center] would import goat hair from India," recalled Plotkin. "And this goat hair was often contaminated with anthrax. Factory workers would get infected when they would inhale anthrax spores from the hair. But Alex [Langmuir] was astonished that someone was willing to go to Philadelphia. He was reminded of the apocryphal tombstone of W. C. Fields, 'Rather here than in Philadelphia.' Unfortunately, I didn't know a damn thing about anthrax. I wanted to go to Philadelphia to work on polio." Going to Philadelphia gave Plotkin a chance to work at the world-famous Wistar Institute. For Plotkin, Wistar would be his McGurk Institute and Hilary Koprowski, the director, his Max Gottlieb.

THE WISTAR INSTITUTE OF ANATOMY AND BIOLOGY, THE OLDEST independent institute for medical research in the United States, sits in the center of the campus of the University of Pennsylvania in West Philadelphia. An impressive three-story brownstone building, the institute, founded in 1892, was named for Caspar Wistar, the nation's leading anatomist in the eighteenth century and the man for whom the plant wisteria was named. Wistar wrote the first textbook on anatomy in the early 1800s and developed a series of anatomical teaching aids that included wax-preserved human limbs and organs. The institute's museum of anatomy contained a cyclops; Siamese

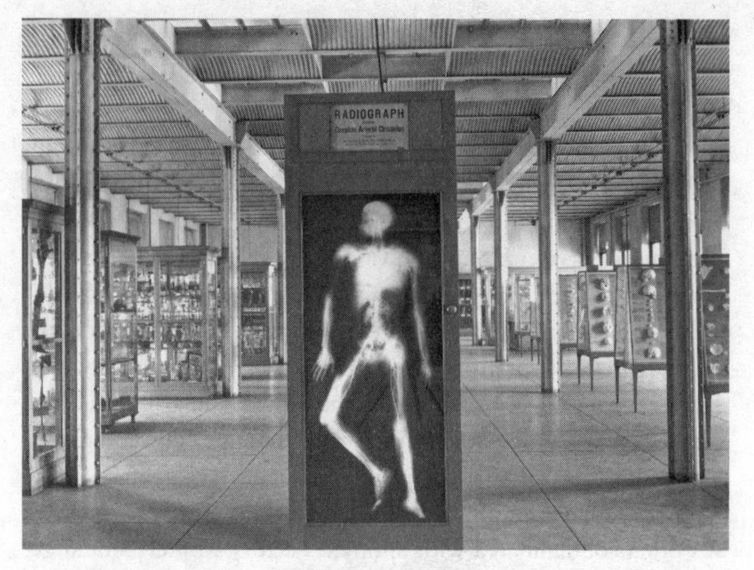


*The Wistar Museum, circa 1940s. An X-ray of a living person injected with a dye to visualize arteries (foreground), was donated by the Eastman Kodak Company in 1939.*

twins; two Indian mummies; seven wax-injected human hearts; the death masks of Oliver Cromwell, Sir Isaac Newton, and Voltaire; and the largest number of human and animal skeletons in the world. A massive whale skeleton hung from the ceiling. Although the Wistar Institute had been founded to advance studies of anatomy, by the early 1960s Hilary Koprowski had made it one of the world's leading institutions for the study of cancer and viruses.

Born in Poland, Koprowski had come to Wistar from Lederle Laboratories in Pearl River, New York. Like the fictional Max Gottlieb, Koprowski had a thick accent and a passion for science. In the 1950s Koprowski was locked in a race with Albert Sabin to make the first polio vaccine using live, weakened human polio viruses. Koprowski was winning the race. He was the first to weaken polio virus by passing it through rats, the first to give the vaccine to children, and the first to test it in thousands of people. But Sabin's vaccine, made by growing polio virus in cells from monkey kidneys and monkey testicles, was eventually judged to be safer than Koprowski's. By the early 1960s, Albert Sabin's polio vaccine would be dropped onto sugar cubes and given to children throughout the world, eliminating polio from many countries. Although he lost his race with Sabin, Koprowski directed a research institute that developed vaccines against rubella, rabies, and rotavirus and advanced our understanding of how and why cells become cancerous.

AFTER COMPLETING HIS STUDIES ON ANTHRAX, PLOTKIN LEFT WISTAR. "By 1961 I had finished my work on polio and anthrax and was looking for something else," he recalled. "So I decided to leave Wistar and complete my pediatric training at the Great Ormond Street Hospital in London. Before leaving, I wrote several grants [seeking funds] to study rubella. During my year in London an outbreak of rubella started in the United Kingdom that was considerable." Working with Alistair Dudgeon, a British virologist, Plotkin saw hundreds of babies permanently harmed by the virus. "It was an experience that you don't get from reading books," he said.

After a year in England, Plotkin returned to the United States. "I was ready to start my own lab at Wistar, focused on rubella, and Hilary Koprowski was very supportive. Lo and behold, the epidemic crossed the Atlantic and swept across the United States, leaving behind thousands of damaged babies: something the popular press was talking about. During the height of the epidemic, 1 percent of all babies in Philadelphia were born to mothers who were infected with rubella." Plotkin was forced to tell hundreds of mothers that rubella virus had probably damaged their unborn fetuses. Many chose to end their pregnancies. "It was a powerful experience," he recalled.

By offering women the option to abort their pregnancies, Plotkin had technically violated the Hippocratic Oath, a pledge that he had taken when he graduated from medical school. In the fourth century B.C., Hippocrates wrote, "I will not give to a woman an abortive remedy." Virtually every medical school in the United States reads the Hippocratic Oath to graduating students, but the oath has been modified since Plotkin took it. Hippocrates' prohibition of abortion has been removed. Also gone are his edict forbidding euthanasia: "I

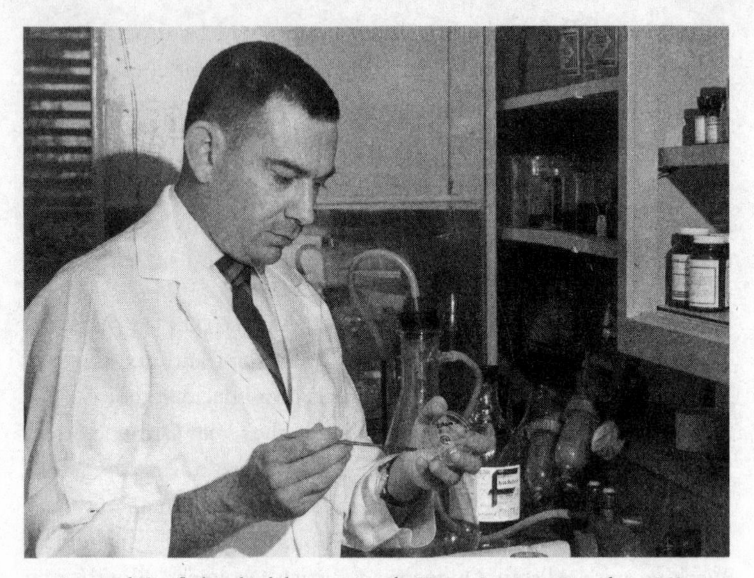

*Leonard Hayflick in his laboratory at the Wistar Institute, circa late 1960s.*

will neither give a deadly drug to anybody who asked for it, nor will I make a suggestion to this effect"; discouraging sexual contact with patients: "Whatever houses I may visit, I will come for the benefit of the sick, remaining free of all mischief, and in particular of sexual relations with both male and female persons, be they free or slaves"; forbidding surgery: "I will not use the knife" (not such a terrible idea in 400 B.C.); and educating medical students for free: "to teach them this art, if they desire to learn it, without fee and covenant."

Plotkin was desperate to make a rubella vaccine. But unlike Meyer and Hilleman, he didn't look for rubella virus in the backs of people's throats. "I looked for rubella in the fetus and not in the throat," recalls Plotkin, "because I couldn't be sure that patients weren't carrying other viruses in their throats, whereas the fetus is in a sterile environment." In 1964, a twenty-five-year-old Philadelphia woman was eight weeks pregnant when she noticed a faint red rash on her face. Fearing that she had rubella, she came to see Plotkin. After confirming her worst fear, Plotkin advised her of the risks to her baby. A few weeks later, the fetus arrived in his laboratory. Because this was the twenty-seventh aborted fetus that Plotkin had received and because he captured rubella virus from the third organ that he tested—the kidney—Plotkin called his vaccine virus *R*ubella *A*bortus 27/3 (RA27/3).

Now that Plotkin had his virus, he had to find a way to weaken it. Whereas Hilleman and Meyer had used animal cells to weaken rubella virus, Plotkin wanted to use fetal cells. He didn't have to look very far to find them. Leonard Hayflick, who briefly shared a lab with Plotkin, was working with cells from a human abortion; he gave Plotkin the fetal cells that he needed to make his rubella vaccine. For his generosity, Hayflick would soon become the principal target of those opposed to using a human fetus to make a vaccine. Initially, Hayflick wasn't particularly interested in vaccines. He wanted to understand how and why we age.

The world's oldest person, Jeanne Calment of France, died in 1997 at the age of 122, having far surpassed the average lifespan in most countries, which is 76. No matter how well we eat, how much we exercise, or how attentive we are to our own safety, we all will die. We die because our cells become less and less capable of doing what they are supposed to do: making needed enzymes, fighting infections, and resisting transformation to cancer. Hayflick wanted to understand why this happened. To do it, he started with the youngest cells he could find.

Leonard Hayflick was a native of west Philadelphia and a graduate of the University of Pennsylvania. His father made dental devices. To get the cells that he needed, Hayflick asked Sven Gard, a Swedish virologist and later member of the Nobel Prize committee, to provide him with a human fetus. "[Gard] had worked in the lab across from me," recalled Hayflick, "and he said, 'We can get you all of the fetal [organs] you want from Stockholm. You know we do this every day there, and it's legal. Let me talk to my colleagues and we'll set up a system." Gard granted Hayflick his wish, sending him a female fetus aborted at the end of the third month of pregnancy. "The [fetus] came from a woman whose husband was a mariner," recalled Hayflick. "He was apparently a drunkard and she didn't want more children. They didn't send the whole fetus. By that time, I knew what [organs] I wanted. I wanted a lung or a kidney. And so they removed these organs [and] put the material in cell-culture fluid, in test tubes or small flasks, packed it in wet ice, and sent it by air, just ordinary airmail."

Hayflick found that fetal cells, placed in laboratory flasks, reproduced. If he took some of these cells and placed them in another flask, they reproduced again. But Hayflick also found that the number of times fetal cells reproduced wasn't infinite. After about fifty doublings, cells would deteriorate and die. "I didn't think very much of it initially," recalled Hayflick. "I just thought everybody knew . . . that normal cells could not divide infinitely; only cancer cells were immortal." Hayflick demonstrated in the laboratory what had been posited by the German biologist August Weissman eighty years earlier: "Death takes place because cell division is not everlasting but finite."

The phenomenon of limited cell doublings is called the Hayflick limit. At first, researchers thought that Hayflick was merely observing a phenomenon unique to cells growing in laboratory flasks and that cells could reproduce forever under the right conditions. But Hayflick found that cells always knew exactly how many times they had reproduced. If he took cells and froze them for months or years, thawed them out, and placed them back in laboratory flasks, they would pick up right where they had left off, doubling a total of fifty times. Hayflick's work seemed to contradict that of Alexis Carrel, the man who kept the chicken heart alive in his laboratory for thirty-two years. But Carrel had unknowingly cheated. While nourishing the heart with crude extracts made from chicken embryos, technicians were inadvertently adding new cells to the culture. "Alexis Carrel was an egomaniac," recalled Len Hayflick. "The technicians knew that chicken cells were contained in the chicken-embryo extracts they were adding to the heart. But they were scared to tell Carrel, worried that it might torpedo his career and that they'd get fired."

In one brilliant experiment, Hayflick proved that it wasn't the conditions of growth that determined how long cells could reproduce. He took female fetal cells that had reproduced ten times and male fetal cells that had reproduced thirty times. Then he put the two groups of cells in the same flask to see how many more times each would divide. Because only male cells had a Y chromosome, he could tell which cells were male and which were female. Hayflick found that under the exact same growth conditions, the female cells divided forty more times, but the male cells only twenty more times. It was as if each cell had an internal clock that dictated how long it was supposed to live.

Later, Hayflick and other researchers figured out why cellular replication isn't infinite. Essential to cell life and growth are the cell's genes, which are encoded on long strands of a chemical called deoxyribonucleic acid (DNA). When cells duplicate, DNA is also duplicated. Responsible for duplicating DNA is an enzyme called DNA polymerase. To begin the process of DNA replication, the polymerase sits on top of DNA as a train sits on top of a track. The train (polymerase) moves forward and reproduces the track (DNA) in front of it but not the small part of the track initially under it. That means the polymerase doesn't reproduce all the DNA. So each time a cell reproduces, its DNA gets a little shorter and shorter. James Watson, who with Francis Crick was the co-discoverer of the structure of DNA, called this the end-replication problem.

But not all cells follow Hayflick's limit and die. Cancer cells, for example, are immortal. They continue to divide over and over again in people and in the laboratory. If, like mortal cells, their DNA is getting shorter and shorter with each round of replication, how do they do that? The answer is an enzyme called telomerase. The end of the DNA that doesn't get reproduced is called a telomere. Cancer cells solve the problem of DNA shortening with telomerase, which goes back and reproduces the tail end of the DNA that the DNA polymerase missed. Scientists are now investigating the role of telomerases in prolonging the life of normal cells, perhaps as a step forward in our hunt for immortality.

HOLLYWOOD MENTIONED THE HAYFLICK LIMIT AND ALSO ADDRESSED the problem of immortality in the 2004 movie *Anacondas: The Hunt for the Blood Orchid*. The story begins in New York City with several pharmaceutical company executives listening to a presentation by research scientists. The scientists describe a rare orchid found only in Borneo: *Perrenia immortalis*, the blood orchid. The orchid, it appears, has miraculous powers. The following exchange takes place:

> Scientist: I don't suppose anybody here is familiar with the Hayflick limit?
> Executive: Hayflick proposed that a cell could only replicate fifty-six times before it died from a harmful build-up of toxins. According to him, it's the reason that we die.
> Scientist: But what if we could transcend that limit?
> Executive: That would be impossible.
> Scientist: Our research indicates a chemical in *Perrenia immortalis* that would significantly prolong life.
> Executive: Does this mean what I think it means? Are we talking about a pharmaceutical equivalent to the Fountain of Youth?
> Executive #2: That would be bigger than Viagra!
> Executive: Well, what the hell are you waiting for? Get your asses down to Borneo.

The researchers go to Borneo and, for the most part, get eaten one by one by incredibly large, mobile anacondas. (Immortality apparently has a price that even Hayflick hadn't considered.)

In the end, Hayflick concluded, "My goal is to live until the age of one hundred and drop dead on my hundredth birthday with full cognitive and physical abilities."

ALTHOUGH STANLEY PLOTKIN SHARED A LABORATORY WITH LEONARD Hayflick, he didn't share Hayflick's interest in understanding immortality. He was interested only in making a vaccine. So Plotkin took cells from Hayflick's aborted fetus and infected them with rubella virus. Rather than grow the virus in cells maintained at normal body temperature—98.6 degrees—Plotkin grew it at 86 degrees. After twenty-five consecutive passages, the virus grew well at lower temperatures but poorly at body temperature. Plotkin had his vaccine. When he tested it in thousands of people, he found that it induced better and longer-lasting protection against rubella than did Hilleman's modification of Harry Meyer's vaccine. (Although we'll never know, it would have been interesting to determine whether Hilleman's original Benoit strain of rubella vaccine—the one he abandoned at the request of Mary Lasker—was comparable to Plotkin's vaccine.)

Plotkin knew that his vaccine worked and that it was safe. But his choice to use human fetuses would soon meet stiff opposition from an unexpected source: Albert Sabin, the man who had beaten Hilary Koprowski to develop the first live, weakened polio vaccine. Like Salk, Sabin was the son of Russian Jewish immigrants, but unlike Salk, Sabin was mean-spirited and occasionally vindictive. By the late 1960s, Sabin was a well-known and influential scientist. His polio vaccine had replaced Salk's and was well on its way to eliminating polio from the Western Hemisphere.

Sabin worried that the fetal cells Plotkin had used to make his rubella vaccine would become cancerous and that they might harbor dangerous human viruses. Plotkin remembers his showdown with Sabin: "In February of 1969, a three-day meeting on rubella vaccines was held on the campus of [the National Institutes of Health] in Bethesda, Maryland. The meeting was attended by a packed house of hundreds of interested parties, including Albert Sabin. Although Albert had not himself worked on rubella, he was there as a guru. As the meeting progressed, I heard that Sabin had made a number of statements in private deprecating the use of [my vaccine]. The last morning of the meeting there was even an interview in the *Washington Post* in which that opinion was stated. Finally, toward the end of the meeting, Sabin rose to inveigh against [my vaccine] in his rabbinical style, darkly alluding to some unknown agent that might be lurking there. But he didn't have any evidence. It sounds theatrical, but I remember that these words from the Bible came into my mind: 'The Lord has delivered him into my hands.' After he sat down, I took the microphone and criticized his statements one by one and at length, pointing out that they were strictly *ex cathedra* and without a factual basis. Much to my surprise, I received a thunderous ovation. The great thing about science is that authority doesn't hold sway. Eventually scientific studies will be the deciding factor and outweigh prevailing opinion. Science is always self-correcting. Today's heresy becomes tomorrow's orthodoxy."

Hilleman later persuaded Merck to replace his modification of Meyer's rubella vaccine with Plotkin's vaccine. "Sometime in 1978, I believe, I was sitting in my office when the phone rang, and it was Maurice Hilleman," recalled Plotkin. "Maurice said that Dorothy Horstmann [a Yale University researcher] had convinced him that it would be a good idea to replace his vaccine with mine. After recovering my faculty of speech, I readily agreed."

During the development of their rubella vaccines, Hilleman and Plotkin shared the same fear: namely, that their vaccines would be given to women who didn't know they were pregnant. Since 1969, thousands of pregnant women have been inadvertently injected with rubella vaccine; many of these women were fully susceptible to rubella and were inoculated during their first trimester of pregnancy. But only one fetus has ever been harmed by the rubella vaccine—probably the single best evidence that both vaccines were remarkably safe.

On March 21, 2005, a dream that in the early 1960s seemed unimaginable came true. The director of the CDC, Dr. Julie Gerberding, held a press conference to "formally and officially declare that rubella has been eliminated from the United States." Many schools for the deaf are now out of business.

It had been little more than sixty years since Norman McAlister Gregg had first noted that rubella virus caused severe birth defects. As of 2021, about 90 percent of the world's countries and territories have used rubella vaccine, all with dramatic results. If the number of countries using the vaccine continues to increase, it's possible that rubella will be eliminated from the world within a hundred years of the discovery of its penchant for attacking unborn children. But for now, rubella virus still harms tens of thousands of unborn babies in the world every year.

DESPITE THE SUCCESS OF THE RUBELLA VACCINE, PLOTKIN'S CHOICE TO use cells from an aborted fetus angered pro-life groups. But Plotkin wasn't the only one to use these cells to make his vaccine. Vaccines against rabies, chickenpox, hepatitis A, and SARS-CoV-2 virus also use them. Before the dust settled, the controversy surrounding fetal cells would lead to official pronouncements by the Catholic Church, rebuttals by prominent researchers, and the persecution of Leonard Hayflick.

# CHAPTER 7

## Political Science

*"We find after years of struggle that we do not take a trip, a trip takes us."*

JOHN STEINBECK, *TRAVELS WITH CHARLEY*

At the front of the opposition to Stanley Plotkin's rubella vaccine stands Debi Vinnedge, director of the Children of God for Life in Largo, Florida: a "pro-life organization focused on the bioethical issues of human cloning, embryonic, and fetal-tissue research." Vinnedge, a mother of two and grandmother of five, is angry that Plotkin used a human fetus to make his vaccine. "Casually accepting the use of aborted fetal cell lines in medical treatments has been a blatant disgrace to humanity," she said, "a despicable sullying of the value and dignity of human life, and has lent credibility to the gross commercialization of aborted babies, ripped from their mother's womb so that someone could turn a profit. We must become slaves to the Culture of Death. Using aborted babies as products to help those children fortunate enough to not have their lives snuffed out pre-birth is akin to the most vile form of cannibalism imaginable. Yet we are asked to accept it for every polite reason except one that begs the question: 'What kind of a civilization have we really progressed to when we can find no better way to protect ourselves than by using the remains of murdered children?'"

Although Vinnedge's words are wildly inflammatory, the logic of those opposed to using a human fetus to make a medical product is clear:

The supreme teaching authority of the Catholic Church—as illuminated by sacred scripture and the teaching of the apostles—is defined in the catechism.

The catechism of the church holds that abortion is intrinsically evil, grave enough to warrant excommunication.

The Vatican and the National Conference of Catholic Bishops have denounced abortion and fetal research.

Therefore, Vinnedge argues, a Catholic, according to his conscience and under the direct teaching of the Catholic Church, has the absolute duty to refuse any medical product derived from an aborted fetus.

Debi Vinnedge knew that Stanley Plotkin's rubella vaccine contained DNA from a human fetus. She simply couldn't allow this DNA to be injected into the arms of children. So on June 4, 2003, she wrote a letter to Cardinal Joseph Ratzinger, then head of the Catholic Church's Congregation of the Doctrine of Faith. Ratzinger was a well-known theologian and prolific author, later to become Pope Benedict XVI, the 265th pope. Today Joseph Ratzinger is Pope Benedict XVI, the 265th and reigning pope.

In July 2005, Vinnedge received a carefully worded response from the Vatican's Pontifical Academy for Life. It wasn't the answer she was looking for. The academy reasoned that those involved with the original abortion had "formally cooperated with evil." Furthermore, those who used the aborted fetus to make vaccines were engaged in an act that was "equally illicit." But the doctors and nurses who give vaccines are engaged in only a "very, very remote" form of cooperation with evil, so remote that "it does not indicate any [negative] moral value" when compared with the greater good of preventing life-threatening infections. The Vatican reasoned that parents who refused Stanley Plotkin's rubella vaccine might be responsible for abortions and damaged fetuses caused by rubella. Such parents would be in "much more proximate cooperation with evil" than if they had accepted a morally questionable vaccine.

The National Catholic Bioethics Center, based in Boston, agreed with the Vatican's decision. "Clearly the use of a vaccine in the present does not cause the one who is immunized to share in the immoral intention or action of those who carried out the abortion in the past. Human history is filled with injustice. Acts of wrongdoing in the past regularly rebound to the benefit of descendents who had no hand in the original crimes. It would be a high standard indeed if we were to require all benefits that we receive in the present to be completely free of every immorality in the past."

But the Vatican didn't let vaccine makers completely off the hook. It emphasized that using an unethical vaccine in no way reflected the church's approval of its production. "The lawfulness of the use of these vaccines should not be misinterpreted as a declaration of the lawfulness of their production, marketing, and use. There remains a moral duty to continue to fight and to employ every lawful means in order to make life difficult for the pharmaceutical industries that act unscrupulously and unethically." The National Catholic Bioethics Center took the Vatican's warning one step further: "The true scandal here is not that Catholics use these vaccines, but that researchers and scientists who bring us these products do not take into sufficient account the moral convictions of millions of their fellow citizens."

Because their experiences with abortion were different, Debi Vinnedge and Stanley Plotkin have strongly opposing views. The Catholic Church taught Vinnedge that a child is created at conception and that abortion is murder. Nothing can justify murder. To Vinnedge, the notion that children could be injected with a vaccine that contained small amounts of DNA from an aborted human fetus was unconscionable. Plotkin (who is Jewish) was an infectious disease specialist during one of the worst rubella epidemics in history. He watched rubella virus kill thousands of babies in their mothers' wombs and cause thousands more to be born blind, deaf, and developmentally disabled. He spoke with hundreds of mothers who came to his office asking whether they should end the life of their unborn children. Plotkin was shaped by what he saw. He decided to do whatever he could to prevent rubella. "Seeing what I saw about the damage that rubella virus could do to infants," he said, "I consider the use of [fetal] cells as 100 percent moral. Frankly, I think that our rubella vaccine has prevented more abortions than all the antiabortionists put together."

Vinnedge doesn't argue with the need to prevent rubella. She argues with the choice of fetal cells to do it. She wonders why Plotkin didn't just use animal cells. After all, Max Theiler used mouse and chicken cells to make his yellow fever vaccine; Jonas Salk and Albert Sabin used monkey cells to make their polio vaccines; and Maurice Hilleman used chicken cells to make his measles, mumps, and pandemic influenza vaccines. All of these researchers made their vaccines by growing human viruses in animal cells. But they were lucky. Some human viruses don't grow very well in animal cells; they grow well only in human cells. Plotkin chose fetal cells to make his vaccine because rubella was one such virus.

But there was another more important reason to use fetal cells: they weren't contaminated with animal viruses. For example, Hilleman found chicken leukemia virus in the measles vaccine that he got from John Enders. Max Theiler's yellow fever vaccine was contaminated with the same virus. But the problem with chicken leukemia virus paled in comparison with what Hilleman found in the late 1950s, when he was interested in making his own polio vaccine—a finding that scared researchers away from animal cells and toward the safe haven of human fetal cells.

ALTHOUGH THEY DIDN'T KNOW IT AT THE TIME, JONAS SALK AND Albert Sabin made polio vaccines that were contaminated with a monkey virus. This particular monkey virus had never been seen before, caused cancer in animals, and had already been injected into millions of children.

To make their vaccines, Salk and Sabin had used kidney cells from rhesus and cynomolgus monkeys, two species that had a long history as research animals. National Aeronautics and Space Administration (NASA) scientists had sent rhesus monkeys into space, and hematologists had used them to define a protein located on the surface of human red blood cells, the Rh factor. Cynomolgus monkeys were also popular among researchers. Behavioral psychologists studied them because they were the only nonhuman primates to wash their food before eating it. And religious leaders worshipped them. Standing in front of the Nikko Toshogo Shrine in Japan are three cynomolgus monkeys, with hands over their eyes, ears, and mouths, representing the religious principle "If we do not hear, see, or speak evil, we ourselves shall be spared evil." Cynomolgus monkeys are the source of the warning to "see no evil, hear no evil, speak no evil."

When Salk and Sabin were making their polio vaccines, thirty-nine different viruses had been found in monkeys. Although regulatory agencies were obviously concerned about these viruses, they were reassured by the fact that monkey viruses infected monkeys much better than people and that all of the viruses were completely killed by formaldehyde. But Hilleman wanted to make vaccines that were completely free of monkey viruses. He didn't want to have to rely on formaldehyde to kill them. So in 1958, while attending a conference in Washington, D.C., he called William Mann, director of the National Zoo. Mann invited Hilleman to his home that evening. "He was a very entertaining man," recalled Hilleman. "I told him about the huge problem that we had with contamination and what it was doing to the whole field of viral vaccines. And he said, 'Come to my home and meet my Montana wife.' [That night] we went into his living room and I thought that I was in a god-damned African tent with these darts and arrows and masks and voodoo dolls. The place was fully decorated with African artifacts." Mann explained why the contamination of monkey cells was so common. "The [monkeys] were trapped in Africa and off-loaded at airports, where they would share their viruses," he said. "They would all be forced together in small spaces, sharing their urine and feces. And employees of the airports, who had no idea of what they were doing, were in charge of feeding and handling the animals. It was a real mess." But Mann also had the solution. "He told me that I had a simple problem," said Hilleman.

"[He] told me to go to West Africa and pick up African green [monkeys], so-called grivets. There are a lot of them there. Bring them into the Madrid airport. No primates other than human beings come to Madrid. Then put them on a big transport airplane, transport them to New York, off-load them there, and you've got it made." Using Mann's African green monkeys, Maurice Hilleman was about to discover the fortieth monkey virus.

Hilleman hired a trapper to catch several African green monkeys from West Africa, paid to have them transported to Madrid, and picked them up when they arrived in New York City. When Hilleman got them back to his laboratory, he killed them, removed and ground up their kidneys, placed the kidney cells into laboratory flasks, and examined them to see if they contained any viruses. First, he looked at them through an electron microscope. Nothing. Then he took the kidney cells, disrupted them, and added them to a variety of other cells to see if any viruses grew. Again nothing. Hilleman was satisfied that kidneys from African green monkeys, if freshly caught and directly transported from West Africa, were free of contaminating monkey viruses.

Then Hilleman performed one more experiment. "There was always the nagging apprehension that a culture [of cells] wasn't free of detectable viruses, but how did I know that there wasn't a virus in there that I wasn't detecting?" Hilleman took kidney cells from rhesus and cynomolgus monkeys that had been deemed free of contaminating viruses—cells that were used routinely to make several vaccines—and added them to his African green monkey cells. Soon the cells developed large holes, clumped together, and died. Hilleman reasoned that a monkey virus was killing them. He called the new virus *S*imian *V*irus 40, or SV40.

Next, Hilleman performed a series of experiments that terrified public health officials and within several years made SV40 one of the best studied viruses in the world. Hilleman injected SV40 into newborn hamsters and saw that in 90 percent of them large tumors developed under the skin as well as tumors in their lungs, kidneys, and brains. Some of the tumors weighed nearly half a pound—more than twice the weight of the hamster. Hilleman then found that even though Jonas Salk had used formaldehyde to make his polio vaccine, his vaccine still contained very small quantities of live SV40. At the time of Hilleman's discovery, Salk's vaccine had already been injected into tens of millions of people, and thousands more were receiving it every day. Hilleman also found that Albert Sabin's polio vaccine, which wasn't treated with formaldehyde, was heavily contaminated with SV40. Sabin's vaccine hadn't been licensed in the United States, but it had been given to ninety million people in Russia, mostly children.

In June 1960 Maurice Hilleman stood at the podium of the Second International Conference on Live Poliovirus Vaccines in Washington, D.C. Risking the wrath of Sabin, whom he knew to be in the audience, Hilleman decided to present his data on SV40. Hilleman explained that he had found SV40 in both Salk's and Sabin's vaccines and that SV40 caused cancer in hamsters. Everyone in the room understood the implications of what he had said. In a paper published later that year, Hilleman stated the obvious. "The lack of significant harmful effect in man, in the short term, is well established in results of studies in millions of volunteers fed [Sabin's vaccine] to date. Less can be said concerning a possible long-term effect." Sabin saw Hilleman's presentation as an attempt to sabotage his vaccine. "I told him that there was no escaping the fact that his vaccine was contaminated with SV40," said Hilleman. "He went crazy, calling me all kinds of names."

During the next few years Hilleman and others performed a series of studies that were largely reassuring. Researchers found that although SV40 caused cancer when it was injected into hamsters, it didn't cause cancer when it was fed to them. Sabin's vaccine was swallowed, not injected. Researchers later found SV40 in the feces of children given Sabin's vaccine, but none of those children developed antibodies to it. Apparently, SV40 just passed through the digestive tract without causing an infection. Researchers also found that although formaldehyde used in the making of Salk's vaccine didn't completely kill SV40, it did decrease infectivity at least ten thousandfold. The quantity of residual SV40 in Salk's vaccine probably wasn't enough to cause cancer. But at that point, no one was sure.

Horrified that children had been injected with a potentially cancer-causing virus, researchers compared cancer rates in children who had received SV40-contaminated polio vaccines with cancer rates in unvaccinated children. Eight years after the tainted vaccines had been given, the cancer incidence was the same in both groups. The same was true fifteen and thirty years later. And it was true for children who had received SV40-contaminated vaccines in the United States, the United Kingdom, Germany, and Sweden. By the mid-1990s public health officials were confident that the inadvertent contamination of polio vaccines with SV40 didn't cause cancer. Then an investigator at the National Cancer Institute in Bethesda, Maryland, found something that reignited the controversy.

MICHELE CARBONE WAS INTERESTED IN FIGURING OUT WHAT CAUSED cancer. So he took cancer cells and studied their genes, hoping that they might provide a clue. Carbone studied unusual cancers of the chest (mesotheliomas), brain (ependymomas), and bone (osteosarcomas). Unlike leukemia, breast cancer, and prostate cancer, the cancers that Carbone studied are very rare. To his surprise, Carbone found one gene that kept turning up in each of the cancers—a gene that was also found in SV40. Carbone knew that SV40 caused cancer in hamsters and that the types of cancers found in hamsters were similar to those he found in people. The implication was clear: SV40-contaminated polio vaccines in the 1950s and early 1960s caused cancer. Carbone reasoned that the earlier studies had been falsely reassuring.

Carbone's findings stimulated another round of studies. This time, researchers looked specifically for Carbone's unusual cancers. They studied hundreds of thousands—instead of just thousands—of people. But the results were the same. Again and again researchers found that people who had received vaccines inadvertently contaminated with SV40 were not at greater risk for cancer. Furthermore, they found that many people who had genetic fragments from SV40 in their cancer cells had never received contaminated polio vaccines. And they found that many adults had antibodies against SV40 in their blood even though they were born well before SV40-contaminated polio vaccines had been given.

Keerti Shah, a professor at the Johns Hopkins School of Public Health, has been studying SV40 virus for more than forty years. When Carbone found genetic remnants of SV40 virus in lung cancers like mesothelioma, Shah also looked but couldn't find any. "There was a big argument in 1998," recalled Shah. "We could not find SV40 in mesotheliomas. Some labs would never find it, like ours. Some labs would always find it, like Carbone's. There was a lot of controversy. People couldn't figure out why this was happening. And there were many recriminations: 'You don't know how to do the assays. You're messing it up.' Then there was a study done by the National Cancer Institute and the Food and Drug Administration. Nine labs all tested mesotheliomas and also tested normal lungs as controls. In that study nobody found any SV40. And today all of the new [researchers] can't find any evidence of SV40 in brain tumors or lymphomas or mesotheliomas." Shah concluded that "SV40 never caused cancer in people."

Toward the end of his life, Hilleman was asked by Merck not to comment publicly about whether there was a link between SV40 and cancer, fearing that his comments might be misunderstood or used against him in pending litigation. But Hilleman was perfectly willing to talk privately. He had no doubt that the monkey virus that he had found contaminating polio vaccines in the mid-1950s didn't cause cancer in people. "I advised [federal regulators] to pull the [polio] vaccine, but in retrospect I think they were right. Those vaccines never caused cancer. And disrupting the [polio vaccine] program would have cost thousands of lives."

The controversy surrounding SV40 contamination of polio vaccines is unlikely to die down soon, having recently triggered a series of lawsuits and the publication of a book, *The Virus and the Vaccine: The True Story of a Cancer-Causing Monkey Virus, Contaminated Polio Vaccine, and the Millions of Americans Exposed* by Debbie Bookchin and Jim Schumacher. Keerti Shah, who was interviewed extensively for the book, was troubled by its conclusion. "It was a well-researched book," he said. "[The authors] did a great job digging up a lot of old materials. But they got the science wrong. And they saw conspiracies everywhere. There were no conspiracies."

Scared by the ordeal of SV40, researchers in the 1970s and 1980s made vaccines using cells from human fetuses instead of animals: vaccines that prevented rabies, chickenpox, and hepatitis A.

TAD WIKTOR WANTED TO MAKE A RABIES VACCINE. LIKE STANLEY Plotkin, Wiktor would make his vaccine under the tutelage of Wistar's director, Hilary Koprowski. Wiktor met Koprowski in Muguga, Kenya, while both were attending a conference on rabies organized by the World Health Organization. A tall, commanding man with the bearing of an army officer, Wiktor, like Koprowski, was Polish. Koprowski was instantly taken by Wiktor's enthusiasm and intelligence and asked him to join the team at Wistar. Wiktor agreed and would stay until his death thirty years later.

HILARY KOPROWSKI WAS IN AFRICA FOR ANOTHER REASON. DRIVEN to prevent the horror of polio—and locked in a competition with Albert Sabin—he wanted to test his live weakened polio vaccine in African children. Later, two journalists blamed Koprowski and his polio vaccine for the AIDS epidemic, calling him the father of AIDS. In 1992 Tom Curtis, an investigative reporter for *Rolling Stone* magazine, wrote an article titled "The Origins of AIDS: A Startling New Theory Attempts to Answer the Question 'Was It an Act of God or an Act of Man?'" Several years later Edward Hooper, an unpaid stringer for the BBC in London, tried to prove Curtis's allegation in his book *The River: A Journey to the Source of HIV and AIDS*. Curtis and Hooper used the following reasoning to explain why Koprowski's polio vaccine was the source of AIDS: chimpanzees are occasionally infected with an HIV-like virus called simian immunodeficiency virus (SIV); the chimp cells that Koprowski used to make his vaccine were contaminated with SIV; SIV fed to African children mutated to HIV. As further proof, they noted that the AIDS epidemic started in central Africa at the same time and in the same place that Koprowski had inoculated children with his polio vaccine.

But Curtis's and Hooper's theories, while fascinating, were flawed. First, the AIDS epidemic didn't begin where Koprowski had performed his studies. Second, Koprowski didn't use chimp cells to make his vaccine; he used monkey cells. (Chimps aren't monkeys; they're apes.) Finally, although some strains of SIV may have been precursors to HIV, mutation from one to the other would have taken decades, not years. The final proof that Koprowski's polio vaccine didn't contain either SIV or HIV came when researchers tested it using a sensitive assay called polymerase chain reaction (PCR), a technique that can detect very small quantities of viral DNA. Not surprisingly, researchers didn't find SIV, HIV, or chimp cell DNA in Koprowski's polio vaccine. In September 2000, Curtis's and Hooper's theories were finally and officially discredited on the stage of the Royal Society of London. Stanley Plotkin, who participated in the African polio vaccine trials, later commented on Edward Hooper's search for the smoking gun behind the AIDS epidemic. "There was no gun, no shooter, no bullet, and no motive."

In May 2006 Beatrice Hahn and her colleagues at the University of Birmingham in Alabama proved that HIV was derived from SIV. Genetic analysis confirmed that SIV, which infected wild chimpanzees in Cameroon, had crossed over to HIV in the early 1930s, twenty years before Koprowski began his polio vaccine trials in Africa. "The genetic similarity was striking," she said. Presumably, a hunter in rural Cameroon was either bitten by a chimp or cut while butchering one.

WIKTOR AND KOPROWSKI WANTED TO IMPROVE ON THE RABIES VACCINE that had been made before them. They knew that because Pasteur's rabies vaccine was made from rabbit spinal cords, it occasionally caused weakness, paralysis, coma, and death. They also knew that researchers in the mid-1950s made a vaccine using duck embryos that was much less likely to cause these side effects. But because the duck vaccine still contained some cells from the duck's brain and spinal cord—which contain myelin basic protein, an occasional

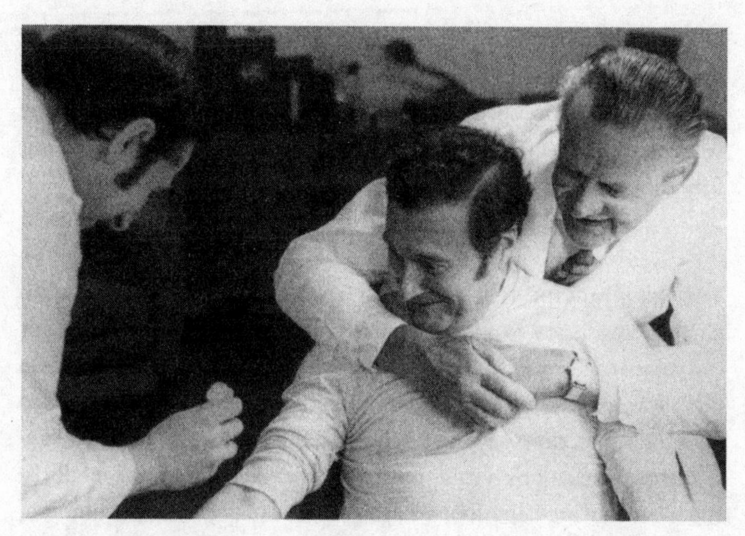

*Stanley Plotkin (left) inoculates Hilary Koprowski with experimental rabies vaccine as Tad Wiktor smiles for the camera, December 1971.*

cause of autoimmunity—it didn't completely solve the problem. Also, the duck vaccine had to be given daily for about three weeks—twenty-three shots in the arms, legs, and abdomen. The procedure was so torturous that many people feared the vaccine more than they feared rabies. To make a better vaccine, Wiktor did what Stanley Plotkin had done. He visited Leonard Hayflick, his colleague at the Wistar Institute, and asked if he could have cells from the abortion that had been performed in Sweden in 1961. Wiktor knew that Hayflick's cells didn't contain SV40, were free of other contaminating viruses, and didn't contain cells from the fetus's brain and spinal cord—exactly what he needed. Within a few years Wiktor found that he could grow rabies virus in Hayflick's cells and completely inactivate it with a chemical. Stanley Plotkin took Wiktor's rabies vaccine and injected it into the arms of Koprowski and Wiktor, finding that it evoked high levels of rabies antibodies. Encouraged, the Wistar team took their vaccine to Iran, where wild rabid dogs roamed the streets, and injected it into people who had been severely bitten. The vaccine was 100 percent effective, worked after only a few injections, and was remarkably safe.

Using Hayflick's fetal cells, Tad Wiktor and Hilary Koprowski had solved the rabies vaccine problem, making a vaccine that is now given to ten million people every year. Unfortunately, many people who need the vaccine still don't get it. Worldwide, rabies kills about fifty thousand people every year.

THE NEXT VACCINE TO BE MADE FROM A HUMAN FETUS PREVENTED chickenpox. Prior to its invention, chickenpox virus infected four million people every year in the United States and a hundred million people worldwide. Although many people consider chickenpox to be a mild infection—a sort of rite of passage through childhood—it's not. Chickenpox virus infects the brain, causing encephalitis; the liver, causing hepatitis; and the lungs, causing fatal pneumonia. (Patsy Mink, a congressional representative from Hawaii, died of chickenpox pneumonia in 2002.) Perhaps the most frightening problem—and the one that has attracted the most attention from the media—is that chickenpox causes a dramatic increase in diseases caused by Group A streptococci, the so-called flesh-eating bacteria. These bacteria invade the skin and muscles through broken chickenpox blisters. Before the chickenpox vaccine became available, chickenpox caused about ten thousand hospitalizations and a hundred deaths every year in the United States alone.

When clinicians realized just how damaging chickenpox could be, they tried to make a vaccine to prevent it. Thomas Weller, who had used cells from a human abortion to perform his Nobel Prize–winning experiment with polio virus, took the first step. In 1951 Weller's five-year-old son, Peter, got chickenpox. Weller broke open one of the blisters, collected the pus, and inoculated it onto cells from a variety of different species and organs. He discovered that human fetal cells worked best, a finding later confirmed by other investigators. "Every known human virus was found to grow in [human fetal cells]," recalled Hayflick. "That's what made them so attractive for vaccines."

In the 1970s Michiaki Takahashi, a microbiologist working at Osaka University in Japan, took the next step. A slightly built, humble man, Takahashi made his vaccine in an unusual way. Like Weller, he took fluid from a blister of a child with chickenpox, a three-year-old boy whose family name was Oka. Then he passed the virus eleven times at low temperature through cells taken from an abortion in Japan, twelve times through fetal cells obtained from guinea pigs, two times through Hayflick's cells, and five more times through cells from a fourteen-week-old male fetus aborted in the United Kingdom in 1966. The resultant vaccine virus is known as the Oka strain.

Hilleman developed Takahashi's vaccine and introduced it into the United States in 1995. By 2005, with almost all children receiving it, the number of infections and deaths from chickenpox declined 90 percent.

THE LAST VACCINE TO BE MADE USING HAYFLICK'S CELLS PREVENTED hepatitis A virus. Before the vaccine, hepatitis A caused about two hundred thousand cases and a hundred deaths every year in the United States. In developing countries, where sewage and drinking water often mix, hepatitis A infects millions and kills thousands every year; almost everyone is infected. Hepatitis A virus also caused one of the largest single-source outbreaks of an infectious disease in United States history. It happened in western Pennsylvania.

Jennifer Seavers was fourteen years old when she went out for Mexican food to celebrate a friend's birthday. Although Jennifer didn't like Mexican food, she didn't want to miss the party. So in October 2003 she and a dozen girls from Ambridge High School sat down to plates of nachos, fajitas, and tacos at a Chi-Chi's restaurant in Beaver, Pennsylvania, twenty-five miles northwest of Pittsburgh. Within a few weeks, one of Jennifer's friends came down with fever, stomachaches, muscle pain, weakness, nausea, and vomiting. Her urine turned dark brown, her skin turned yellow, and she couldn't breathe without feeling as if she was being stabbed just below her ribs. Frantic, the girl's parents took her to the doctor, whose blood test revealed the diagnosis: hepatitis A.

Jennifer knew that she had dodged a bullet. "I feel really lucky that I didn't get sick," she said. "But I know a lot of people who did." Jennifer's friend was one of many. The first case of hepatitis occurred on October 2, 2003. During the next few weeks, several more people in that area got sick. By November 3, local health officials confirmed that there was an outbreak of hepatitis and asked Chi-Chi's to shut its doors. Because the time from exposure to the virus to the appearance of first symptoms can be as long as seven weeks, everyone who had eaten at the restaurant from September to November was at risk; Chi-Chi's had served eleven thousand meals during that time. By November 5, the number of people infected had risen to eighty-four; by November 6, to a hundred and thirty; and by November 7, to two hundred. On November 7, Jeffrey Cook, an auto-body restorer from Aliquippa, Pennsylvania, died following a liver transplant in a desperate attempt to save his life.

Health officials were sure that the outbreak was caused by a restaurant employee, and during their investigations they found several employees with hepatitis. But all had been infected at the same time as their customers, not before. The timing wasn't right. The virus was coming from somewhere else. By November 11, the number of people infected had risen to three hundred, and two more people were in critical condition. By November 12, after forty more people became ill, investigators finally closed in on a likely source. Earlier that year, public health officials had interviewed thousands of people during outbreaks of hepatitis A in Georgia, Tennessee, and North Carolina. They had asked people who had and hadn't gotten sick what they had eaten. Then they had carefully checked the ingredients of each of those foods. One item kept appearing on the list of hepatitis A victims: green onions (scallions). Investigators found that restaurants had imported their green onions from Mexico, a country with a high rate of hepatitis A infection. Richard Quartarone, a spokesman for the Georgia State Health Department, said, "Because they're multilayered, green onions are very difficult to clean. The only way to be 100 percent sure that you've killed the hepatitis in the green onions is to cook them. But they're often used as a garnish, so you don't cook them." Chi-Chi's, which had imported its green onions from Mexico, immediately pulled them from the menus of their ninety-nine other restaurants from Minnesota to the mid-Atlantic. But it was too late.

By November 13, the number of cases had risen to four hundred, and the outbreak claimed its second victim, Dineen Wieczorek, a customer service representative for Ikea. She had eaten at Chi-Chi's on October 6 to celebrate her wedding anniversary. Her daughter, Darleen Tronzo, recalled, "One meal. One meal, that's all it took. And people eat out every day. I eat out every day. You never think something like this could come of it." By November 14 the number of cases had risen to five hundred, and the outbreak claimed its third victim, John Spratt, an employee at a payroll processing company. Spratt had eaten at Chi-Chi's with his daughter. They had both ordered the chicken fajitas. But Spratt, not his daughter, had chosen to eat the condiments that came with the meal—a choice that killed him.

When it was over, the hepatitis A virus outbreak at Chi-Chi's had infected about seven hundred people and killed four. Although this was the largest outbreak of hepatitis A in the United States, it wasn't the largest outbreak in the world. In 1989 in Shanghai, hepatitis A virus in uncooked polluted clams from the East China Sea had infected more than three hundred thousand people and killed forty-seven. Clams, like mussels and oysters, can filter as much as ten gallons of water an hour, concentrating the virus a hundredfold relative to its concentration in the ocean.

At the time of the outbreak at Chi-Chi's, a vaccine that could have prevented the tragedy had been available for eight years. Critical to the development of the first hepatitis A vaccine was Friedrich "Fritz" Deinhardt.

In the mid-1960s, Fritz Deinhardt was an unknown researcher. Born in Gütersloh, Germany, he attended the University of Göttingen and received his medical degree from the University of Hamburg. After completing both an internship and a residency in Hamburg, Deinhardt came to the United States, where he met his wife, Jean, and later became the chairman of microbiology at the Presbyterian St. Luke's Hospital in Chicago. In 1965, while working at St. Luke's, Deinhardt sampled the blood of a thirty-four-year-old surgeon whose initials were G. B. The surgeon had been ill with hepatitis for three days, his skin and eyes had turned yellow, he couldn't eat without vomiting, and he was tired and listless, unable to work. To capture the surgeon's virus, Deinhardt took the blood and injected it into the veins of white-lipped, hairy-faced marmosets—small, squirrel-like monkeys found in South America. Within a few years, marmosets would become an endangered species. But before the United States government prohibited research on marmosets, Deinhardt imported them, bred them, and, with the help of his wife—who fed milk to the babies—raised them. A few weeks after Deinhardt had injected marmosets with the surgeon's blood, all of the animals fell ill with hepatitis.

Despite his apparent success at isolating a hepatitis A virus, the American military, which funded Deinhardt's studies, was skeptical of his results. They asked Hilleman what he thought of Deinhardt's work. Although Deinhardt could be difficult to deal with—aggressive and argumentative—Hilleman stood by his friend, recalling that "[Fritz] took samples of blood from a surgeon with the initials G. B. and isolated a virus from that. He reportedly had hepatitis A. [Deinhardt] showed he had transmitted the virus to marmosets. All the time that the military was funding that, they would ask me, 'Do you think that he's doing anything? Do you think he's right?' I said 'Fritz Deinhardt is very bright. And who in the hell are you going to get to do the hepatitis A work if Fritz doesn't do it? You've got to fund him.' I had no reason to believe at the time that he hadn't isolated hepatitis A virus."

On April 30, 1992, at the age of sixty-six, Fritz Deinhardt died of cancer. One of Deinhardt's obituaries, in reference to his work on marmosets, stated, "It was here that the first roots of meaningful specific studies of hepatitis A were begun. This breakthrough discovery was the epicenter of all hepatitis A investigations, opening a window to that which followed, and initiating a trail to the reality of [a] vaccine. Were it not for these seminal marmoset findings, we might still be struggling with the mysteries of hepatitis A." But Fritz Deinhardt had never studied hepatitis A virus. He had isolated a very rare virus now called hepatitis G virus, an uncommon cause of disease in humans. Although he was unknowingly studying a different virus, Deinhardt was right about the marmosets; they were an excellent model for the study of hepatitis A.

Following Deinhardt's lead, Hilleman injected blood from a nine-year-old Costa Rican boy with hepatitis into the vein of a marmoset. Several weeks later he detected hepatitis A virus in the marmoset's liver. But marmosets were becoming increasingly harder to find. He needed other cells in which to grow his virus. And Hayflick's cells were the only ones that worked.

During the next thirteen years Maurice Hilleman became the first person to detect hepatitis A virus and hepatitis A antibodies, the first to grow the virus in fetal cells, the first to weaken it (as an added margin of safety), and the first to kill it with formaldehyde. Then he became the first to show that his weakened-then-killed hepatitis A vaccine worked in animals. Confident, he was ready to test it in people. When he tested his measles, mumps, and rubella vaccines, Hilleman had needed people at high risk for those diseases, so he had chosen institutionalized, developmentally disabled children. Now he needed people at high risk for hepatitis A infection. Fortuitously, at around the time that Hilleman was looking for a place to do his studies, Alan Werzberger visited Merck. Werzberger was a physician at the Kiryas Joel Institute of Medicine.

In 1974, to accommodate their growing population in the Williamsburg section of Brooklyn, a group of Hasidic Jews moved to the Hudson Valley, fifty miles northwest of New York City. The village that they established, Kiryas Joel, transformed a quiet rural setting into a bustling urban center, with bearded men wearing black coats on hot summer days and women in head scarves pushing baby carriages past signs written in Hebrew. They brought a unique way of life from Brooklyn. Inhabitants of Kiryas Joel married at eighteen and had large families. By the early 1990s, about eight thousand people lived there.

The village of Kiryas Joel stood out for yet another reason: it had unusually high rates of hepatitis A. Typically, when infants and young children are infected with hepatitis A virus, they don't suffer any symptoms of the infection. But when older children, teenagers, and adults are infected, they often suffer severe disease. At

*Hasidic boy living at Kiryas Joel receives an experimental hepatitis A vaccine while Alan Werzberger (left), who conducted the clinical trials, looks on, 1991.*

Kiryas Joel, the living conditions allowed for easy transmission of the virus from younger children to older children. "The high birth rate, large family size, [and] day care center–like school atmosphere permitted close contact between younger and older children," said Werzberger. "[We had] difficulty maintaining constant control over what the [younger] children did with their hands, from peremptory hand washing to surreptitious dips into communal school food." "They all bathed together in these community pools," recalled Phil Provost, Hilleman's co-worker on the hepatitis A vaccine at Merck. "It was a traditional practice in the community. But there was a lot of hepatitis A virus contamination in those pools. That's how they were spreading it." From 1985 to 1991 local physicians cared for three hundred children infected with hepatitis A virus at Kiryas Joel. The virus infected 70 percent of all residents.

The task of testing Hilleman's weakened-then-killed hepatitis A vaccine fell to Alan Werzberger. First, Werzberger found a thousand children who had never been infected with the virus. Then he divided them in half; one half received a shot of vaccine, the other a shot of placebo. Three months later he found that hepatitis A virus had infected thirty-four children in the study, all of whom had received placebo. In a paper published in the *New England Journal of Medicine*, Werzberger concluded that Hilleman's hepatitis A vaccine was 100 percent effective.

Since 1995, when Merck licensed its hepatitis A vaccine, the incidence of the disease in the United States has declined about 75 percent.

When Stanley Plotkin, Tad Wiktor, Michiaki Takahashi, and Maurice Hilleman made their vaccines using Hayflick's cells, they didn't consider their actions to be immoral. Fetal cells were attractive because they were free of contaminating animal viruses, easy to grow in the laboratory, and highly susceptible to every known human virus. They were, in many ways, ideal cells in which to make vaccines. At the time they were being made, no one complained: not the media, the public, religious groups, the FDA, the NIH, or the WHO. Also, the woman who had provided her fetus to Leonard Hayflick for use in vaccines had requested the abortion. But times have changed. Now some groups—like those headed by Debi Vinnedge—see vaccines made from fetal cells as immoral.

Vinnedge sees a simple solution. Now that we have more sophisticated methods to detect contaminating viruses, we can remake these vaccines in animal cells. It wouldn't be easy. First, companies would have to find the right animal cells in which to grow these viruses. Then they would have to effectively weaken or kill the viruses, test these new vaccines in progressively larger trials (including tens of thousands of children), construct or refit buildings in which to make the vaccines, solicit FDA approval in the United States, and solicit approval from other regulatory agencies throughout the world. Because these diseases are now uncommon, it would be difficult or impossible to conduct trials large enough to prove that they worked. And because these new vaccines might not work as well, these trials could be considered unethical. Furthermore, from the point of view of regulatory agencies, these companies would be making new products that would have to be subjected to the standard scrutiny of a new product. Regulatory burdens would be immense; each new vaccine would cost at least $800 million to make. These new vaccines wouldn't increase sales, just costs. And these costs would have to be absorbed by taxpayers, medical insurance premiums, and international health agencies.

It's unlikely that vaccine makers are going to remake routine children's vaccines—such as those for rubella, hepatitis A, and chickenpox—at great cost for no financial benefit. And inflammatory, incorrect statements regarding vaccines in current use don't help. "The widespread use of these vaccines makes it very likely [that] additional cell lines will be created from other aborted babies," says Vinnedge. "As opposed to the Nazi Holocaust, the abortion Holocaust is ongoing." But the truth is that no new abortions are performed to make any of these vaccines. The cells frozen from the abortion performed in 1961, periodically thawed and grown in laboratory flasks, constitute all that is necessary to make them for generations.

LEONARD HAYFLICK, THE MAN WHO HAD GIVEN FETAL CELLS TO PLOTkin, Wiktor, Hilleman, and Takahashi to make their vaccines, became a pariah. In 1968 Hayflick left the Wistar Institute to become a professor of medical microbiology at Stanford University School of Medicine. He took his fetal cells with him. Hayflick had already set up a company called Cell Associates, with him and his wife as sole proprietors, which sold fetal cells to hundreds of researchers throughout the world. Hayflick charged researchers his costs for preparing and shipping the cells, never profiting from the sales; the total proceeds were only about $15,000. But some saw Hayflick as profiting from his work. "In those days, in that environment," recalled a Wistar co-worker, "when you did research with government support, it was in the public domain. When it came out that Len was selling these cells, a lot of people were appalled."

Well funded by the NIH, in the midst of making important discoveries about how and why we age, and well respected by many of his fellow scientists, Hayflick was at the top of his game. But two years later Leonard Hayflick was standing in an unemployment line in Palo Alto.

On January 30, 1976, James W. Schriver, a management accountant at the NIH, filed a report claiming that Hayflick was selling something that wasn't his to sell. Schriver claimed that because Hayflick's research was funded by the NIH, money made from the sale of fetal cells should be paid to NIH, not Hayflick. Stanford's School of Medicine, alarmed by the growing scandal, concluded that Hayflick had acted unethically. On February 27, 1976, Leonard Hayflick resigned. "In 1975 I took the first initiative and asked the director of NIH to make a definitive determination about [who owned my fetal] cells," recalled Hayflick. "Instead of sending a lawyer or a scientist, who might understand my claim, they sent an accountant, who went to Clayton Rich [Stanford's dean] and said 'Do you know that you have a thief in the microbiology department?' [The dean was] advised by the bookkeeper that his cooperation was expected because 90 percent of [Stanford's] budget came from NIH. Stanford University called the campus police and asked them to call the district attorney. Public servants from the NIH [then] entered my laboratory, confiscated [my fetal cells], and claimed them for themselves. The NIH maintained that it was perfectly fair for commercial organizations, the Russians, and themselves to sell [my fetal cells] for tens of millions of dollars, but they viewed as theft the inventor or his institution profiting." In the eyes of his fellow scientists, Hayflick was ruined. "I went from full professor at Stanford to the unemployment line in one week," he recalled. "My wife and I lived on $104 a week for the next year." Plotkin remembered the controversy: "I think that in the really classical Greek sense it was a tragedy because here was a man who at the height of his powers brought about his own downfall."

Hayflick sued the federal government, which in turn countersued. "I felt, and I think I was justified in feeling, that these cells were like my children," he said. Hilleman, who was asked to testify against Hayflick, recalled, "I was asked to be a principal witness against him. And I said that if there was an attempt to convict him, I would make a campaign on my part that two top-level government officials would spend time in jail with him. He should have been celebrated as a scientific hero instead of being persecuted." In September 1982, after six years of wrangling, the case was settled out of court—in Hayflick's favor. Hayflick was awarded the principal plus interest of the $15,000 that had been held in escrow, and the government allowed him to keep his cells. Hayflick never kept the settlement award, using it to pay his lawyers. Colleagues rallied to his support. A letter signed by eighty-five scientists was published in the journal *Science*: "This happy outcome of Dr. Hayflick's courageous, sometimes lonely, emotionally damaging, and professionally destructive ordeal provides several important object lessons for the future. In light of the settlement terms and other government actions, few will disagree that the original allegation against him was entirely unjustified." Hayflick's battle changed the law. Now scientists receiving federal money can own and sell their discoveries. This single ruling allowed for the boom in private-sector biotechnology in the 1980s and 1990s. "I was a pioneer," recalled Hayflick. "And it's the pioneers that have the arrows in their backs."

Although the use of fetal cells to make vaccines remains controversial for some, the vaccines made from them are safe. Fetal cells allowed Hilleman and others to avoid contaminating viruses like chicken leukemia virus and SV40. But Maurice Hilleman was about to use a material to make his next vaccine that few thought was safe, even after the vaccine had been licensed and sold: human blood. Hilleman obtained blood from drug abusers and homosexual men living in New York City in the late 1970s, when HIV first entered the United States. It was arguably the most dangerous starting material ever used to make a medical product.

# CHAPTER 8

# Blood

*"We had a process that would destroy all life forms."*

MAURICE HILLEMAN

In 1984 researchers at the CDC published a paper titled "Cluster of Cases of the Acquired Immune Deficiency Syndrome (AIDS): Patients Linked by Sexual Contact." AIDS, a syndrome that included unusual infections and cancers, was sweeping across the country. Thousands had been infected.

Victims of AIDS died of many different diseases. For example, they died of pneumonia. Before AIDS entered the United States, pneumonia caused by pneumococcal bacteria killed tens of thousands of people every year. But AIDS patients were different; they were killed by *Pneumocystis*, an organism previously found to cause pneumonia only in cancer patients. They also died of meningitis—but again, not from typical bacteria, such as meningococci, but from unusual fungi such as *Cryptococcus*. Or they died of Kaposi's sarcoma, a previously rare form of cancer that caused hideous dark purple spots under the skin.

The CDC researchers found several groups of people at high risk for AIDS: Haitians living in the United States, intravenous drug users, and people who required frequent blood transfusions. But no group was at greater risk than homosexual men. The first forty people diagnosed in the United States with AIDS were gay men living in California, Florida, Georgia, New Jersey, Pennsylvania, and Texas. To figure out how the AIDS virus—soon to be called human immunodeficiency virus (HIV)—spread, investigators constructed a diagram showing who had had sex with whom. In the center of the diagram was one man. All forty AIDS victims had had sex with this man or with someone who had had sex with him. They called him Patient Zero. His name was Gaetan Dugas.

Born and raised in Quebec, Dugas, a steward for Air Canada, traveled extensively throughout the United States, frequenting many gay bars and bathhouses. When he walked into bars, he would stand in the entrance, scan the room, look carefully at each patron, and declare, "I am the prettiest one." And he was. Randy Shilts in *And the Band Played On* described Dugas as having "sandy hair that fell boyishly over his forehead, an inviting smile, and a laugh that could flood color into a room of black and white." His sexual escapades were legendary. "In San Francisco," wrote Shilts, "Gaetan returned from every stroll down Castro Street [the center of the gay community] with a pocketful of matchbook covers and napkins that were crowded with addresses and phone numbers. At times [he] would stare at his address book with genuine curiosity, trying to recall who this or that person was."

Dugas was twenty-eight years old when a biopsy of an enlarging purple spot below his right ear revealed Kaposi's sarcoma—"gay cancer." At the time, Dugas estimated that he had slept with two hundred and fifty men a year for ten years—twenty-five hundred sexual partners in all. Knowing that AIDS was contagious didn't stop Dugas from continuing to satisfy his sexual appetite. "Rumors began on Castro Street about a strange guy at the Eighth and Howard bathhouse, a blond with a French accent," noted Shilts. "He would have sex with you, turn up the lights in the cubicle, and point out his Kaposi's sarcoma lesions. 'I've got cancer,' he said. 'I'm going to die. [And now] so are you.'"

. . . .

A FEW YEARS BEFORE HIV FIRST ENTERED THE UNITED STATES, Maurice Hilleman began working on a new vaccine—not for HIV, which was still unknown, but for hepatitis. Because of the method he chose for making it, fear of AIDS would soon spread to fear of Hilleman's vaccine. For almost two hundred years, researchers had used cells from monkeys, chickens, mice, rabbits, and ducks to make their vaccines. Hilleman was about to break new ground. He would be the first (and last) to use human blood to make a vaccine. He didn't know until later that the blood was heavily contaminated with HIV.

Several viruses infect the liver. But by far the most common, the most severe, and the most feared is hepatitis B virus, which infects one third of the world's population, about two billion people. Most people infected with hepatitis B virus recover completely. But not everyone recovers. Some people die of an overwhelming infection in a matter of weeks. Others have a persistent infection: one million people in the United States and more than three hundred million people in the world are chronically infected with hepatitis B virus. And most of them don't know it. Victims of chronic hepatitis B are at high risk for two possible fates: dying of cirrhosis, a progressive destruction of the liver, or dying of liver cancer. Hepatitis B virus is the third most common known cause of cancer in the world. The sun, which causes skin cancer, is the first; cigarette smoking, which causes lung cancer, is the second.

BEFORE MAURICE HILLEMAN COULD MAKE HIS HEPATITIS B VACCINE, he had to capture the virus. When Hilleman had wanted to make vaccines against measles, mumps, or rubella, he simply swabbed the throats of children with those diseases. Unfortunately, hepatitis B virus is barely detectable in saliva. Blood, on the other hand, contains extraordinarily large quantities of the virus—about five hundred million infectious particles per teaspoon. But blood from people infected with hepatitis B contains more than just virus particles. It contains something that will ultimately lead to the eradication of hepatitis B virus.

Every virus has a different strategy for survival. To avoid provoking an immune response that would destroy them, chickenpox and herpes simplex viruses live silently, latently, in the nerves. Many people initially recover from infection only to have these viruses reemerge decades later in the form of shingles or herpes blisters. Influenza virus outsmarts the immune system by constantly changing one of its surface proteins, the hemagglutinin. People make antibodies to the hemagglutinins of influenza viruses one year, only to find that these antibodies don't completely protect them the following year. So influenza virus continues to thrive. Rabies virus, which lives in saliva, evades the immune system entirely. After entering the body through the bite of an infected animal, it slowly, inexorably travels up the nerves of the arm or leg to the brain, moving from one nerve cell to the next, never entering the bloodstream. Many people infected with rabies virus make rabies antibodies. But by traveling from cell to cell, the virus effectively hides from antibodies in the blood. When rabies virus finally reaches the brain—a process that takes about two months but can take as long as six years—death is inevitable.

HIV is probably the most heinous because it infects one particular group of cells: T cells, which are important in directing the immune system. When T cells are destroyed, the immune system is disabled. Worse, HIV evolves rapidly during infection; people make antibodies to the virus only to find that different HIV viruses have taken the place of the old ones.

Hepatitis B virus has a strategy for survival that is different from that of any other known virus. In order for hepatitis B virus to infect the liver, it must first bind to liver cells via a protein that sits on the surface of the virus. People make antibodies against this viral surface protein to prevent the virus from attaching. If the virus can't bind to liver cells, it can't infect them. But hepatitis B virus fights back by making far more viral surface protein than it needs to make new virus particles, hoping that this excess surface protein will soak up antibodies from the blood and allow free virus to attach to liver cells. Hepatitis B virus is so committed to this method of survival that people infected with the virus have about five hundred quadrillion (500,000,000,000,000,000) particles of viral surface protein circulating in their bodies during infection. But hepatitis B virus's strategy of overproducing surface protein would eventually prove to be its Achilles' heel.

To make his hepatitis B vaccine, Hilleman followed a trail blazed by Baruch Blumberg, a researcher working at the Fox Chase Cancer Center in northwest Philadelphia. Blumberg wasn't a virologist, an immunologist, or an infectious disease specialist. He was a geneticist. For the longest time, while studying hepatitis B surface protein, he didn't have the faintest idea of what he was looking at.

A stocky, powerful, outgoing native of New York City, Baruch Blumberg got a degree in physics from Union College in Schenectady, New York, before attending the College of Physicians and Surgeons at Columbia University. The single event that changed his life occurred in the early 1950s between his third and fourth years of medical school. "Harold Brown, our professor of parasitology," recalled Blumberg, "arranged for me to spend several months at Moengo, an isolated mining town accessible only by river, in the swamp and high bush country of northern Suriname [in South America]." Moengo was a melting pot inhabited by Javanese, Africans, and Chinese, as well as Hindus from India and Jews from Brazil. Blumberg found that people with different backgrounds had different susceptibilities to certain infections.

One infection common in Suriname was elephantiasis, caused by *Wuchereria bancrofti*, a tiny worm that blocks the flow of lymphatic fluid from the legs or genitals, causing massive, disfiguring swelling. Legs become coarse and thick, and scrotums become so swollen that victims have to carry them around in wheelbarrows.

*Wuchereria bancrofti* causes severe elephantiasis in some people but mild or no disease in others. Blumberg found that susceptibility to diseases like elephantiasis could be directly linked to ancestry. He reasoned that people with different susceptibilities to a particular disease made proteins that served the same function—for example, to counteract the disease—but that these proteins were slightly different in their size or shape. They were thus known as *polymorphisms*, for "many forms." Researchers had already found several protein polymorphisms. For example, they found that people have different proteins—A, B, and O—on the surface of their red blood cells. Blood group protein differences are important. If a person with type A blood receives type B blood, that person will make antibodies to the type B protein that destroy the transfused cells. The reaction can be massive and fatal. This is why doctors determine a patient's blood type before transfusion.

Blumberg's hypothesis that disease susceptibility is genetic is probably best shown by the origin and function of one specific type of hemoglobin, called hemoglobin S. Hemoglobin, a protein found inside red blood cells, also has several different forms. Fetuses and newborn babies have hemoglobin type F; most children and adults have hemoglobin type A; and some people, mostly of African descent, have hemoglobin type S. These three different hemoglobin proteins have the same function—to carry oxygen from the lungs to the rest of the body—but they are clearly different in size and shape. It isn't a coincidence that hemoglobin S is found mainly in people from Africa. People with hemoglobin S are better able to resist malaria—a parasite common in Africa—than are those with hemoglobin A. When the malaria parasites enter red blood cells, hemoglobin S eventually causes cells to change shape, making it more difficult for the parasites to survive. Unfortunately, some hemoglobin S–containing red blood cells, which look like tiny sickles, have difficulty passing through small blood vessels. The genetic adaptation to malaria infection is called sickle cell disease.

Looking for protein polymorphisms, Blumberg examined blood from people who had received at least twenty-five blood transfusions. He reasoned that people receiving many blood transfusions would be more likely to have antibodies to proteins different from their own. In 1963 Blumberg found that the blood from a man with hemophilia in New York City contained antibodies to a protein found in the blood from someone halfway across the world, an Australian Aborigine. He called the protein in the Aborigine's blood Australia antigen. (An antigen is a protein that evokes an immune response.) Blumberg found that Australia antigen was very rare in the United States—only one of every thousand people had it—but that it was quite common in tropical and Asian countries.

At this point, Blumberg didn't know what he had stumbled upon. Two years later, in 1965, Blumberg found to his surprise that Australia antigen was common in people with leukemia. He thought that the protein was either a marker for leukemia or part of a virus that caused leukemia. By 1967 he had found that Australia antigen, in addition to its presence in people with leukemia, was often present in the blood of Americans with Down syndrome. Again he thought that because children with Down syndrome were at higher risk for leukemia, Australia antigen was a marker for leukemia. But children with Down syndrome were more likely to have Australia antigen in their blood because they were more likely to have been infected with hepatitis B virus, the result of living in places like Willowbrook. Blumberg still hadn't realized that he had discovered a protein that was part of hepatitis B virus.

Eventually a virologist named Alfred Prince, working at a transfusion center in New York City, figured it out. In the early 1960s, Prince took blood from people before and after they received transfusions. In 1968, he found a patient who had hepatitis. Early samples of the patient's blood didn't contain Blumberg's Australia antigen, but later samples did. Prince concluded that "[Australia] antigen is located on a virus particle and the virus particle is etiologically related to some or all cases of serum hepatitis [soon to be called hepatitis B virus]." Prince was the first person to realize that Australia antigen was part of hepatitis B virus. Ten years later, in 1976, Baruch Blumberg won the Nobel Prize in medicine for discovering Australia antigen. His acceptance speech mentioned Alfred Prince briefly, parenthetically, and unfairly: "The Australia antigen association was also confirmed in 1968 by Dr. Alberto Vierucci, who had worked in our laboratory, and [that of] Dr. Alfred M. Prince."

Now that researchers knew that Australia antigen was a protein made by hepatitis B virus, they could begin to investigate the possibility of using it to make a vaccine. Saul Krugman, the infectious diseases specialist who had fed hepatitis virus to developmentally disabled children at Willowbrook, performed the controversial experiment.

Krugman was born in the Bronx, the son of Russian immigrants. His family later moved to Paterson, New Jersey, near the home of Krugman's first cousin Albert Sabin. In high school, Krugman was a lively, outgoing member of the debate team, the drama club, and the student council. After high school he attended Ohio State University until he could no longer afford it, dropping out after his junior year. He worked for a year, finally graduating from the University of Richmond and later the Medical College of Virginia. Two years later, during the Second World War, he served as a flight surgeon in the South Pacific, earning a Bronze Star. When the war ended, Krugman returned to New York and took a position at the Willard Parker Hospital as an extern (an intern without salary). From this lowly rank he made a steady climb, eventually becoming a professor of pediatrics at the New York University School of Medicine and chairman of the department of pediatrics. Krugman was a latecomer to academic medicine. He didn't publish his first scientific paper until he was thirty-nine years old. Nevertheless, by the end of his career he had published two hundred and fifty more papers and was the coauthor of a leading textbook in infectious diseases, now in its eleventh edition. Krugman's colleagues remember him as honest, thoughtful, hardworking, and highly ethical: a wonderful father, mentor, and friend. But his hepatitis B experiments, although they paved the way toward prevention of the disease, later caused many in the media and the public to see him as a monster.

Knowing the work of Blumberg and Prince, Krugman took blood from a patient with hepatitis B virus infection, let it clot, took the serum, and injected it into the veins of twenty-five developmentally disabled children, again at Willowbrook. He wanted to see whether serum from people with hepatitis contained hepatitis virus. Not surprisingly, it did. Twenty-four of the twenty-five patients became sick as the virus attacked their livers. Krugman concluded that "this study indicated that serum was highly infectious for susceptible individuals." One of the children sickened by the injection of live dangerous hepatitis B virus was still infected five years later, likely eventually to have either cirrhosis or liver cancer.

Now that Krugman had found an infectious serum that made children sick, he wanted to see if he could use it to protect them. So he took the infectious serum, diluted it in water, and heated it for one minute. Krugman hoped that by heating serum just below the boiling point he would kill hepatitis B virus but leave Australia antigen—hepatitis B surface protein—intact. He gave some children one dose of the vaccine and others two. Krugman then injected these children with untreated infectious serum, knowing that if the vaccine didn't work, virtually all would be infected with hepatitis B virus. The vaccine worked, protecting all of the children given two doses and half of the children given one dose. "It was a very, very exciting time," remembered Krugman. "But I really wasn't trying to develop a vaccine. Actually, all we did in our little laboratory, our little kitchen, so to speak, was [to] boil hepatitis B serum and water."

A local politician soon tempered Krugman's excitement. On January 10, 1967, Seymour Thaler, a New York state senator, took the floor of the senate chambers in Albany. Thaler said that "he had searched his soul and conscience" and that "the medical profession has presumed to act as God over the health and lives of the medically indigent." "I have the documentary proof," he said. "I have undergone a terrible inner conflict on whether to bring to the attention of the public that thousands of patients are being used daily as medical guinea pigs." Jack Hammond, director of Willowbrook, stood up to disagree: "We're not using the youngsters because they are [developmentally disabled], but because hepatitis is a particular problem at Willowbrook. We have the consent of the parents of every child enrolled in the program." New York State medical officials supported Hammond, pointing out that because of Saul Krugman, hepatitis had been virtually eliminated from the school. Thaler wasn't impressed. He introduced a bill banning medical research on children. Although the bill died in deliberation, the effect of the bill and its publicity on Saul Krugman didn't. "That business with Senator Thaler was very difficult," recalled Krugman. "It happened when he ran for reelection. Politicians have to get publicity, so he invited the press to join him when he went to Willowbrook School and held a press conference and [made accusations] completely, of course, out of context. It was difficult."

Saul Krugman's studies at Willowbrook showed that there were two different types of hepatitis virus (A and B), that gamma globulin could prevent disease, and that Australia antigen could be used as a vaccine. Humanity clearly benefited from his work. For his efforts, Krugman received many awards, including the John Howland Award, the Bristol Award, and the Lasker Award. He was also elected to the National Academy of Sciences, one of the greatest honors that a scientist can receive from his peers. And he received a special citation from parents at Willowbrook for having helped their children. But in 1972, when Krugman received an award from the American College of Physicians in Philadelphia, he needed a police escort to protect him from nearly two hundred people who came to denounce him. Protesters, sickened that he had injected disabled children with a dangerous virus, would follow Saul Krugman for the rest of his life.

KRUGMAN KNEW THAT HIS EXPERIMENT AT WILLOWBROOK WAS ONLY the first step. "I don't like to call it a vaccine," he said, "because it really wasn't a vaccine. Our finding was serendipitous. It demonstrated that a vaccine might indeed be developed. What was needed then was for the vaccine manufacturers, with their highly sophisticated technology, to follow up our lead." Blumberg had found Australia antigen. Prince had shown that Australia antigen was hepatitis B surface protein. And Krugman had shown that antibodies to the surface protein protected children against hepatitis B virus infections. "Now all the bells were ringing," recalled Hilleman. "Because all a vaccinologist needs is an antigen. I had to find out [first] whether or not blood from hepatitis B virus carriers had enough Australia antigen [for commercial use] and, second, whether that blood could remain safe."

In the late 1970s, to get enough hepatitis B surface protein for his vaccine, Hilleman sought out homosexual men and drug users, groups at the highest risk of hepatitis B infection. (Many of these people lived in flophouses, stairwells, doorways, and fire escapes in the Bowery, one of the New York City's most notorious neighborhoods.) Then he began a seemingly impossible task. Hilleman took blood loaded with hepatitis B surface protein, live dangerous hepatitis B virus, large quantities of other blood proteins and—unknown to Hilleman at the time—HIV, and purified it so that only hepatitis B surface protein remained. He had no previous studies to guide him, no precedent for this kind of work.

Initially Hilleman decided, as had Krugman before him, to heat the blood. "The program went off in two different directions," recalled Hilleman, "one of which I called Klink's Clunk. Klink was an engineer here at Merck. I told him I wanted him to build a Clunk. And that would be a continuous flow system into which we would put highly purified hepatitis B plasma, and we would pass it through a pipe with hot water, then past an ultraviolet light, and [then] into a pool of formaldehyde. That damn thing was so technical because we had to have constant flow: if we were going to put it into hot oil, everything had to be [processed very quickly]. But Klink's Clunk really hadn't significantly materialized before we developed a chemical process."

Klink's Clunk failed, so it was on to plan B. Hilleman decided to use three different chemicals to treat the blood. He started with pepsin, an enzyme that breaks down proteins. Hilleman hoped to destroy blood proteins, such as gamma globulin, that were present in large quantities, but he didn't want to destroy the hepatitis B surface protein. It worked. "For some reason pepsin didn't destroy [Australia] antigen," recalled Hilleman; "in fact the stuff that came out was almost totally pure." Hilleman found that pepsin reduced the number of infectious hepatitis B virus particles in blood one hundred-thousandfold. But he knew that such a reduction might not be enough to destroy every last virus particle. So he added a second step: urea. A product of protein metabolism, urea is present in large quantities in the urine of mammals (hence its name). Like pepsin, concentrated urea also destroys proteins. Hilleman used urea because it destroyed prions, a particular group of proteins that might also be present in human blood and that were dangerous to humans.

In the mid-1950s a researcher named Carleton Gadjusek traveled to New Guinea to study kuru, a disease characterized by a slow but relentless decent into dementia. Gadjusek found that the disease occurred mostly in cannibals who ate human brains. At first, Gadjusek and others thought that the disease was either genetic or caused by a virus. But neither theory was correct. Kuru was caused by unusual proteins called proteinaceous infectious particles, or prions. Mad cow disease, caused by eating meat contaminated with infected brains or spinal cords, is also caused by prions. And Creutzfeldt-Jakob disease, a similar ailment, although not caused by eating contaminated meat, is also caused by prions. When Hilleman made his hepatitis B vaccine, he was afraid that blood might be contaminated with prions. "I'll tell you one thing that really worried me at the time, and that was Creutzfeldt-Jakob," recalled Hilleman. "This disease was known to be infectious. It had been shown, quite uniquely, that urea would destroy [prions]. We used [urea] and had a pretty good reason to believe that there would be no problem."

Hilleman wasn't finished. He wanted to add one more chemical to destroy any contaminating viruses. So he picked the one that had been used successfully by Jonas Salk to kill polio virus: formaldehyde. Polio virus, like hepatitis B virus, was difficult to destroy, but formaldehyde readily destroyed both.

Now Hilleman had his method: he would treat human blood with a combination of pepsin, urea, and formaldehyde. He knew that each of these treatments caused a one hundred-thousandfold decrease in hepatitis B virus; the combination of the three caused a quadrillionfold (1,000,000,000,000,000) decrease. Hilleman didn't know whether his method would destroy all other contaminating viruses, so he carefully tested representatives of every virus known: viruses similar or identical to rabies, polio, influenza, measles, mumps, smallpox, herpes, and the common cold. These viruses cause infections of the brain, spinal cord, liver, lungs, nose, throat, and intestines. Hilleman's chemical treatments completely destroyed every one of them. "[I thought] that if we could show that each step could kill surrogate viruses," recalled Hilleman, "a whole bunch of different viruses, then we had a process in which viruses would be deader than deader than dead. A process that would destroy all life forms."

As it turned out, hepatitis B surface protein was quite hardy. While the combination of chemical treatments destroyed other proteins in blood like gamma globulin, the hepatitis surface protein remained intact. Hilleman used a series of filtration steps to further purify his vaccine. In the end, Hilleman's blood-derived hepatitis B vaccine was virtually 100 percent pure hepatitis B surface protein—a technical miracle. But the road to a final product wasn't easy. "This was a precedent," recalled Hilleman, "and it really took a lot of god-damned guts to take Merck down that trail. You can imagine the progress that we made. It was just about a damned zero data base [to start]. Shall we drop it this week, or should we wait another week? We were walking around in the dark or in the mud. This was one big gamble, I'm telling you."

In the late 1970s Hilleman didn't test his chemical inactivation method to determine whether it killed HIV because HIV hadn't been discovered yet. Harvey Alter, a microbiologist who worked with Baruch Blumberg, remembered that "Hilleman was very careful about making the vaccine. He inactivated it in many more ways than were absolutely necessary. As it turned out, it was a great thing because then AIDS came along and scared everybody to death about taking vaccines. But he had done all the right things to kill the AIDS virus, even if he didn't know it was in there."

Hilleman had to convince people that a vaccine made from the blood of intravenous drug users and homosexual men was safe. "You can go ahead and [take a protein]," he said, "and you can purify it, and you think you have a way of inactivating it, but you're still pretty much in the area of faith. You still don't know the potency, the safety, or the efficacy." Hilleman had trouble getting permission from the FDA to test his new product. Then he ran headlong into the same man who had tried to torpedo Stanley Plotkin's rubella vaccine, Albert Sabin, a respected virologist whose opinions continued to influence researchers and regulators. "We had to get [permission from the FDA], and we had trouble," recalled Hilleman. "What do you think happened? Sabin hears about it, says that [our] vaccine will not be used in any human being. Sabin said that if there was a lawsuit, he would go to court to testify against us. He would sue Krugman [his first cousin] if there were any problems with the studies. My feeling was, 'Screw you, Albert.' We went to see John Seal [at the NIH]. He advised us against using our [blood-derived] vaccine. He said, 'You know Albert says he's going to go public.' We had futzed around for about one year. I told Saul [Krugman] that I couldn't wait any longer, that I was going to go ahead and put this into people."

Hilleman knew that he would have a tough time convincing people to try his vaccine. So he turned to the one group he was certain would take it—midlevel executives in his own company. "I went to a meeting for marketing, sales, production, and research," recalled Hilleman, "and I headed up the meeting. And I said, 'Look, guys, our next product is going to be a hepatitis B vaccine, but I need to have volunteers.' I said that I could not use lab people because if any of us came down with hepatitis B, that would be the end of the product. I said, 'Here are the consent forms. Just sign these and I'll collect them after the meeting, and then I'll figure out who are the chosen people.'" Hilleman soon found that he hadn't been very persuasive. "There wasn't a damn one of them that sent in the form," he said. At the next meeting Hilleman made it clear that the consent form didn't contain "No" as an option. "I said, 'I need volunteers, damn it. Just decide who among you are going to take this vaccine. Give yourselves a little bit of time to regain your senses.'" Joan Staub, one of those asked to take the vaccine, remembers things differently. "Consent forms? What consent forms?" she asked. "We got that vaccine because we had to get it. If Hilleman told you to do something, you did it." Staub learned months later where the blood had come from and that there was a possibility that it might be contaminated with HIV. "We were scared to death," she recalled. "I thought I was going to die. Maurice pulled us all into one room and had to explain to us over and over again about the inactivation process and that we were going to be OK."

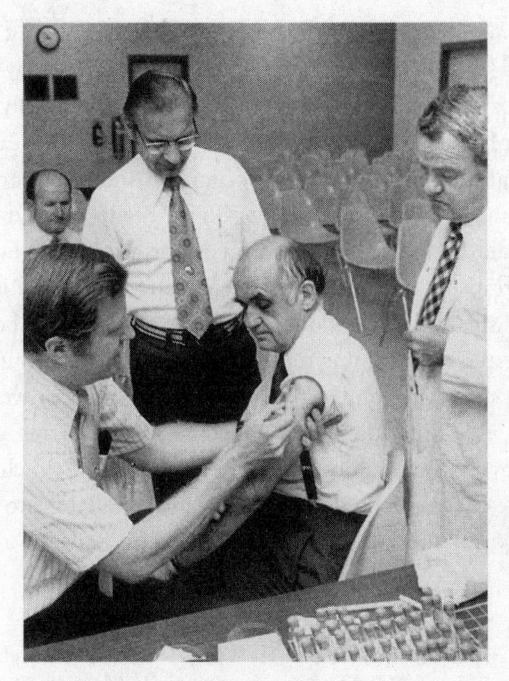

*Robert Weibel injects Maurice Hilleman with an experimental hepatitis B vaccine made from human blood, circa late 1970s.*

Hilleman's coercion of Merck employees to get his hepatitis B vaccine was just one example of his tough, profane, and demanding style. "We worked hard, seven days a week," recalled Hilleman. "If I ever caught anybody delaying a set of tests because [results] might come out on a weekend, it would be grounds for dismissal. You can imagine how that went over. They all had wives and that sort of thing. Now Merck tells [employees] that they don't need to put in any extra time and that you have to balance your life. And that you have to have enjoyment with your job; [that way] you can do a better job and have fun. It's all just a pile of shit. What the company should be doing is kicking ass. But that's from the old school. I was told that I had a very unusual management style."

Hilleman demanded from others what he demanded from himself. "He was totally fixated on what he did," recalled Bert Peltier, former vice president of medical affairs at Merck. "He didn't play cards or have any particular hobbies. He didn't take a lot of vacations. He'd hit the office, and he was just totally at it all day long. He didn't relax very much at all. He was absolutely and totally dedicated to what he was doing. And he could be intimidating. He tended to run roughshod over people below him. He didn't wait around for all of the niceties. He just wanted to do it." Staub recalled that Hilleman would occasionally show disdain for scientists whose work he didn't respect. "Maurice would have different people in to give seminars," she said. "Every once in a while we would all be invited. Mostly people were coming in to pitch something to Maurice, like a program that they wanted to have us fund. One day we were all assembled in a large room, and Hilleman wasn't buying what this guy was saying. Before long I heard a little click, click, click going on a couple of rows behind me. What Maurice had done to show his boredom was that he was back there cutting his nails."

NOTHING WAS MORE INTIMIDATING ABOUT HILLEMAN THAN HIS relentless profanity. "He liked to curse," recalled former Merck CEO Roy Vagelos. "He had language that characterized his being, and he brought that wherever he went. And he never changed." Hilleman recalled, "I remember that at about age three, I was sitting on the kitchen table while [Aunt Edith] was putting on my long stockings. I had the sudden urge to practice my gradually increasing vocabulary. And I said, 'Oh, fuck.' Swat. And I was lying prone on my side. 'Wow,' I said. 'What a powerful word. I wonder what it means.'"

Although Hilleman's daughters Jeryl and Kirsten were never victims of his profanity, they occasionally heard cursing around the house and picked it up. "Dad had enrolled me in a Quaker school about fifteen minutes from our house," recalled Jeryl, "where I had a lovely, dedicated, but very fundamentalist first grade teacher. She proceeded to go on with a lesson about Adam and Eve, and I was sitting in the back of the class. I can remember this as if it were yesterday. I eagerly raised my hand because I understood that she was on the wrong track and really was in need of some of the enlightenment that I had received in my own home. So she looks back and calls, 'Jeryl.' And I proceed to look her in the eye and say, 'My dad says that's a lot of crap.' The next scene in my mind is of this very large, suddenly gigantic teacher coming back, tearing to the back of the room, grabbing me by the arm, pulling me into the bathroom, yanking the soap out of the crank dispenser, and washing out my mouth. I was completely mystified because there was absolutely nothing wrong with the language. I was, after all, helping her to understand the way the world worked. But I really would have loved to have witnessed the call that night to my father who, I'm sure, was hard pressed not to laugh."

By the mid–1960s, hoping to make better, happier workers, Merck hired several psychologists to help senior executives with their management techniques. "About seven or eight years [after I came to Merck], they were sending everybody off to charm school, everybody in upper management," recalled Hilleman. "The purpose was to teach people to get along, to make joint decisions, not to blow up, to suppress the ego. [Sessions] were run by psychologists." Max Tishler had the unenviable task of asking Hilleman to go to these group sessions. (Tishler couldn't possibly have been paid enough money to do this.) "Max said that I had to go to charm school and that I shouldn't cuss so much," recalled Hilleman. "[Max] heard me [curse] at a meeting [of department heads], and he said that I should never use [those words] again. I told him, 'Fuck it. What is this, fucking parochial school?'" Tishler later said, "Usually when people came into my office they left shaking. But when Maurice came into my office, I'm the guy who was shaking." Hilleman never attended the sessions.

Although Hilleman's tirades were legendary, one meeting stands apart from all the rest, recorded in a memo still discussed among Merck employees twenty-five years later. They call it "the truck driver memo." Hilleman's explosion resulted from the confluence of several events. When Hilleman was making his hepatitis B vaccine, he was adamant that his method of inactivation be followed precisely. If not, he feared that children might be injected with live deadly hepatitis B virus. But Hilleman's research group didn't make the vaccine; Merck's manufacturing division did. And the manufacturing division was controlled by the unions: the Teamsters Union (hence truck drivers) and the Oil, Chemical, and Atomic Workers Union. Hilleman, who demanded total control, didn't control the employees who made his vaccines. Roy Vagelos remembers the conflict between Hilleman and the manufacturing division. "[Hilleman] was from Walter Reed, which is military. All you had to do was walk into his [laboratory] to realize that he had transformed [the department of] virus and cell biology at Merck into a military organization. Everybody knew what to do at the beginning of every day. Well, that worked in virus and cell biology, but when he started to transfer the process to manufacturing, the people [there] weren't quite used to that. And so there was a constant rumble at the end of the campus, where the transfer of technology took place." On August 15, 1980, the rumble turned into a roar.

Hilleman found out that someone in the manufacturing division—in hopes of increasing production—had slightly changed his chemical inactivation process for hepatitis B vaccine. Hilleman knew that no test existed for detecting very small quantities of live hepatitis B virus, that there was no safety net if a modified process didn't kill every single infectious hepatitis particle. And he knew that children would be at risk. So he gathered the manufacturers together in a small, non–air-conditioned room at the far end of Merck's West Point campus and told them what he thought about their idea. "There is no fucking test for absolute safety except to put the vaccine in fucking man," said Hilleman. "A procedure was developed to make the fucking vaccine and was shown to make the vaccine safe. Then there are always fucking people who want to make fucking brownie points by changing the process to get more yield. You have to adhere to the goddamn process. We know that the vaccine is safe, but you have to adhere to the goddamn process. What worries me is that [someone] will get a bonus if he can get more yield, so he changes the fucking process. Goddamn meatheads are everywhere." Hilleman knew that for biological products like vaccines the manufacturing process was everything. And he refused to tolerate any change in a process that he knew was perfect, even if a change might mean greater yields, increased profits, and shorter timelines for production. He was obsessed about the safety of his vaccine. And he was intolerant of those whose work ethic was less stringent than his (which was pretty much everybody). "Maurice always wanted to have very tight control over everything that he did," recalled Peltier. "He tolerated fools terribly."

If employees didn't meet his rigorous demands, Hilleman fired them, later lining up their shrunken heads like trophies behind his desk. "One of my favorite gifts was a shrunken head kit," recalled Maurice's younger daughter, Kirsten. "It involved carving apples; inserting assorted grotesque teeth, eyes, and hair; and allowing them to dehydrate [for] several weeks. My dad saw an apparent remarkable resemblance in one of my shrunken heads to an employee he had recently terminated. He carted it off to work and installed it in the cabinet behind his desk. Finding this enormously funny, he put me to work carving the heads of his most memorable terminations. My mother caught wind of this enterprise and, horrified, shut down my production facility."

Despite his iron hand and frightening manner, Hilleman's coworkers were fiercely devoted to him. "As afraid as I was of him, there is nothing that I would ever let tarnish his image," remembers Staub, "because I was so overwhelmed by everything that he did. And there was a point where being overseen by a dictator paid off. There was a day called Black Friday in the 1970s where Merck laid off people, a lot of people. I had never seen Merck lay off people until that day. Not one person in virus and cell biology was touched, and that was because of Maurice. We were afraid of him. There was no doubt. But he protected us." Hilleman successfully lobbied for yearly 10 percent increases in his research and development budgets, even though revenues from Merck's vaccines paled in comparison with those from its drugs. Staub believes that Hilleman's style will never be duplicated. "Today decisions are reached by achieving consensus among committee members," she said. "Maurice had a committee. It was a committee of one. We did what the man wanted, and we did it how he wanted. He was a man of his time. If he showed up at Merck today, he couldn't do it."

. . . .

As far as he knew, Hilleman had successfully purified hepatitis B surface protein from the blood of homosexual men and drug users infected with the virus. Confident that he had his vaccine, he eventually persuaded Saul Krugman, Krugman's wife Sylvia, and nine Merck executives to take it. Krugman supervised the experiment. "I had the nurse give me the first injection," recalled Krugman. "Then I gave everybody else theirs." For the next six months a physician examined each of the volunteers and periodically checked their blood to make sure that they didn't have hepatitis. When the experiment was over, all slowly exhaled.

To test his hepatitis vaccine in more people, Hilleman needed someone whose reputation was unassailable. He chose Wolf Szmuness.

Small, fifty years old, with a pockmarked face and a startling shock of blond hair, Szmuness first came to the United States in 1969. Born in Warsaw, Poland, Szmuness lost his parents in the Holocaust and later fled to Russia ahead of an advancing German army. For twenty years he practiced medicine in Siberia and Russia. One event changed the direction of his career: his wife, Maya, almost died of hepatitis following a blood transfusion. Szmuness spent the rest of his life trying to understand the disease and its causes. In 1959 he returned to the small bustling city of Lublin on Poland's eastern border. But eight years later, in 1967, Szmuness again faced anti-Semitism. During the Six-Day War between Israel and Egypt, government officials ordered Szmuness to protest Israel's actions at a rally. He refused and was fired the next day.

Szmuness gradually worked his way up from lecturer to professor at Columbia's School of Public Health. By 1973 he was directing hepatitis testing for the gay community at the New York Blood Center. An epidemiologist, clinician, public health official, and hepatitis virus researcher, Szmuness was the perfect choice to direct Hilleman's vaccine trial. He quickly recruited a thousand homosexual men who had never been infected with hepatitis B virus. By October 1979 Szmuness had injected half of them with three doses of hepatitis vaccine and half with placebo. In June 1980, almost two years after the trial started, Szmuness and his colleagues carefully analyzed the results. Men who got the vaccine were 75 percent less likely to get hepatitis than were men who didn't get the vaccine. Hilleman's hepatitis B vaccine worked.

This controversial trial would later cause some to blame Hilleman for the AIDS epidemic in America. Although it was clear that AIDS had entered the United States well before Hilleman's hepatitis B vaccine trials, Alan Cantwell Jr., a dermatologist, said in his 1988 book *AIDS and the Doctors of Death*, "To my surprise, I quickly discovered that much of the scientific knowledge that has accumulated on the spread of AIDS in America has come from the surveillance and blood testing of large groups of gay and bisexual men who volunteered as human test subjects in the original hepatitis B vaccine trials, which took place in six American cities during the years 1978–1981. Was it coincidental that these were the beginning years of the new mystery disease in gays, and the years just before AIDS became 'official'?" Cantwell reasoned that Hilleman's hepatitis B vaccine contained HIV and was therefore responsible for spreading AIDS in America. Later, researchers tested Cantwell's hypothesis by examining the incidence of AIDS in men who had or hadn't received Hilleman's hepatitis B vaccine. There was no difference. Although HIV was likely present in the blood from which he had made early preparations of his vaccine, Hilleman's choice of pepsin, urea, and formaldehyde had completely destroyed it. (The publisher of Alan Cantwell's book, Aries Rising Press, was founded by Cantwell himself to promote his uninformed views of the origin of the AIDS epidemic.)

BEFORE MAURICE HILLEMAN COULD SELL HEPATITIS B VACCINE TO THE public, he had one more obstacle to overcome: Baruch Blumberg. While at Fox Chase, Blumberg had patented "a vaccine against viral hepatitis essentially consisting of Australia antigen," U.S. patent number 3,636,191. The patent stated that the vaccine should be made by a process that "comprises substantially removing impurities including infectious components," that there should be a step "to attenuate any virus that might remain," and that the vaccine should be "free of blood components other than Australia antigen." Patents aren't usually issued for ideas; they're issued for specific methods. Blumberg, like everyone else in the field, knew that the hepatitis B vaccine shouldn't contain live hepatitis B virus, that it shouldn't contain any other viruses, and that it shouldn't contain other blood proteins. But he didn't know how to do it. Hilleman saw Blumberg's patent as a lot of handwaving. "People in the hepatitis field were aghast at the guts of this son of a bitch," recalled Hilleman. "Somebody had actually issued a patent for that crap."

Hilleman feared that Blumberg's patent might stand in the way of Merck's vaccine. So he went to the Fox Chase Cancer Center to talk to him. In a spacious room overlooking the wooded grounds of Fox Chase, Hilleman sat uncomfortably among a group of administrators who knew little about viruses or vaccines. Blumberg wasn't in the room. "We needed to go out and talk to [the Fox Chase] people because we wanted this for an international product," recalled Hilleman. "So we went out there and said that we would like to manufacture this product. But we didn't want any interference on sales and marketing. A financial officer speaking on behalf of Fox Chase said, 'Here are the conditions. Blumberg will be the director of the program.' So I went over to the blackboard and I drew a picture." It was a picture of Hilleman submerged in a lake. "Under your proposal this is me," said Hilleman. Then he drew a stone, tied it around his neck, pointed to the figure, and said, "I'm under the water drowning, and that's a millstone, the Blumberg millstone." Hilleman was furious. He wasn't going to give up control of the manufacture and testing of his vaccine—especially to someone who he felt knew little about viruses or immunology. He continued to attack the Fox Chase administrators. "If you're so dumb as to think that somebody who knows absolutely nothing [about viruses] is going to be the director of a program in which we're going to invest hundreds of millions of dollars, you're crazy. I said, 'Divest yourselves of that idea.'" The Fox Chase ad-

*Mary Lasker poses with Lasker Award winners Maurice Hilleman (center, back) and Saul Krugman (back, second from right); both won the award for their work on the blood-derived hepatitis B vaccine, November 16, 1983 (courtesy of the Bettmann Archives).*

ministrators relented, licensed the patent to Merck, and abandoned the idea of having Blumberg run the program.

Years later, when other companies started to make their own blood-derived hepatitis B vaccines, Hilleman showed them the patent that he had licensed from Fox Chase. He assumed that Merck still had exclusive rights. "They laughed like hell," remembered Hilleman. "'You call that a patent?' they asked. 'You paid them money for that?' If I had it to do all over again, I'd wish I'd never gone out to Fox Chase. But I'm one of those guys who don't want war. I'd rather pay somebody a reasonable amount. I felt that [Blumberg] deserved some remuneration for having discovered [Australia antigen]. I thought that his finding was really significant. It opened a door. But that's as far as it went."

In his book *Hepatitis B: The Hunt for a Killer Virus*, Baruch Blumberg claimed that the hepatitis B vaccine was his invention. Maurice Hilleman's name is mentioned once: "In 1975, Fox Chase Cancer Center licensed Merck to develop the vaccine. Dr. Maurice

Hilleman, who had considerable experience in the development and manufacture of vaccines and in hepatitis research, was the executive in charge of the program." Blumberg failed to mention that it was Hilleman who had figured out how to inactivate hepatitis B virus, how to kill all other possible contaminating viruses, how to completely remove every other protein found in human blood, and how to do all of this while retaining the structural integrity of the surface protein. Blumberg had identified Australia antigen, an important first step. But all of the other steps—the ones critical to making a vaccine—belonged to Hilleman. Later, Hilleman recalled, "I think that [Blumberg] deserves a lot of credit, but he doesn't want to give credit to the other guy."

Maurice Hilleman's blood-derived hepatitis B vaccine, licensed by the FDA in 1981, was on the market until 1986, but doctors were reluctant to use it. They remained concerned about the source of the blood, unconvinced by the science. "When we brought [the vaccine] onto the market, we had one hell of a time," recalled Hilleman. "The doctors and the nurses did not want to be vaccinated with human blood." Hilleman knew that his method of inactivation killed all known human viruses. But he also knew that asking physicians to understand the science of viral inactivation was asking a lot. "Chemicals are chemicals," he said. "It doesn't matter where the blood comes from. But it took a really enlightened person to understand the story."

Ironically, Hilleman's blood-derived hepatitis B vaccine, made from the most dangerous starting material ever used, was probably the safest, purest vaccine ever made.

But because clinicians in the United States were uncomfortable using a vaccine made from human blood, Hilleman had to find another way to make it. (Although Hilleman's blood-derived hepatitis B vaccine is no longer made by manufacturers in North America or Europe, it is still made by several companies in Asia.) Fortunately, in the early 1970s, two researchers eating lunch at a delicatessen in Hawaii struck a deal that gave Hilleman the technology he needed to make another hepatitis B vaccine. Its creation would also help to usher in the age of genetic engineering.

. . .

HERBERT BOYER AND STANLEY COHEN STARTED A REVOLUTION IN biology.

Boyer was born in Derry, a dark, industrialized corner of western Pennsylvania best known for its mines, railroads, and quarterbacks; Jim Kelly, Joe Namath, Johnny Unitas, and Joe Montana all played high school football in western Pennsylvania. Boyer also played football as an offensive lineman. But Boyer's football coach was also his science teacher, and his coach's passion for science influenced Boyer more than his passion for football. After high school, Boyer studied biology and chemistry at St. Vincent's College in nearby Latrobe, Pennsylvania, followed by graduate studies at the University of Pittsburgh and postgraduate work at Yale. Then he traveled west, arriving in San Francisco at the height of the 1960s counterculture. With a broad round face, impish smile, thick walrus-like mustache, and a wardrobe of leather vests, blue jeans, and wide, open-collared shirts, Herbert Boyer looked like the rock musician Jerry Garcia from the Grateful Dead. And, like Garcia, he was active in the civil rights movement and vigorous in his protests against the war in Vietnam.

But Boyer had come to California to pursue his love of science, not the counterculture. He took a job as an assistant professor of biochemistry at the University of California in San Francisco. By 1969 *Escherichia coli*, or *E. coli*, a common intestinal bacterium, had caught his attention. Boyer found that *E. coli* made enzymes that neatly and specifically cut DNA. In 1972, while he was in the middle of these studies, Boyer traveled to Honolulu for a scientific meeting. There he met Stanley Cohen, a scientist from Stanford who was also working on bacteria. Cohen had found that some bacteria resisted the killing effects of antibiotics, while others didn't, and that bacteria could transfer this resistance to bacteria living next to them. Then Cohen discovered how they did it. Bacteria carried the genes for antibiotic resistance on small circular pieces of DNA that he named plasmids. Plasmids were promiscuous, easily moving from one bacterial cell to another.

At the Hawaii conference, Boyer and Cohen were each intrigued by the other's work. They decided to meet later that evening. Sitting over corned beef and pastrami sandwiches, they had an idea for how their research could be combined. To test it, they performed an experiment that would generate four hundred product licenses from the FDA, form the basis of fourteen hundred biotechnology companies, and launch an industry with annual revenues of $40 billion. Cohen took Boyer's DNA-cutting proteins, cut a plasmid DNA that contained one antibiotic-resistance gene, and inserted a gene that resisted a different antibiotic. Then the two researchers repaired the plasmid so that it again formed a circle. Now the plasmid had genes that resisted two antibiotics. Cohen reinserted this new plasmid into a bacterium and found that they had created a new bacterium that could now resist killing by both antibiotics. Boyer and Cohen reasoned that any gene, even human genes, could be inserted into bacterial plasmids. Every time these genetically engineered bacteria reproduced and made their own proteins, they would also be making human proteins; bacteria could become tiny factories that mass-produced a wide variety of human products. The new field of research launched by Boyer and Cohen was called recombinant DNA technology, or genetic engineering.

The value of this invention wasn't lost on a venture capitalist named Robert Swanson, who called Boyer and asked to meet him in a San Francisco bar. There, the twenty-nine-year-old Swanson and forty-year-old Boyer drank beer and discussed the commercial value of synthesizing human proteins in a laboratory. On paper napkins, they sketched out plans for the first biotechnology company based on genetic engineering. Boyer named it Genentech, a contraction of *gen*etic *en*gineering *tech*nology. When Genentech went public in 1980, the stock had the most dramatic escalation in the history of Wall Street, raising more than $38 million in capital and making multimillionaires of its founders. Later that year, Boyer's picture was on the cover of *Time* magazine under the heading "Shaping Life in the Lab: The Boom in Genetic Engineering." Genentech's first product was human insulin. No longer did insulin have to be purified from the pancreas of cows and pigs; it could be made by

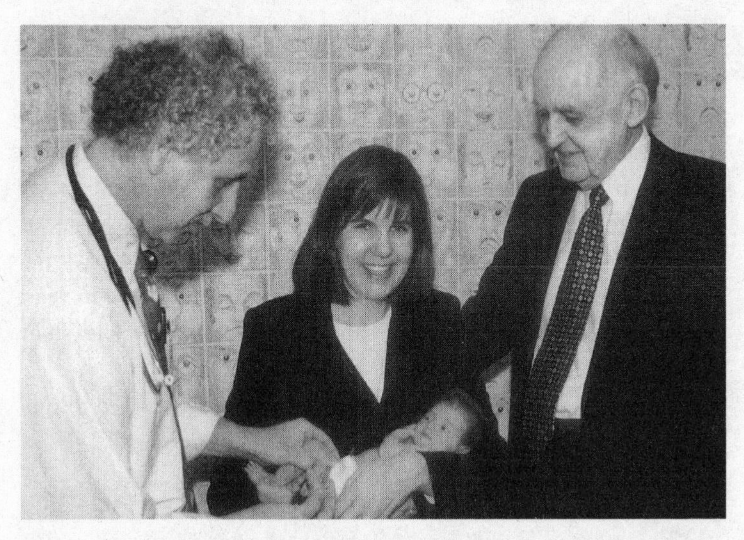

*Dr. Michael Traister (left) inoculates Kirsten Hilleman's ten-day-old daughter, Anneliese, with the recombinant hepatitis B vaccine, 1999. Maurice Hilleman looks on.*

bacteria in a laboratory. Later Genentech made proteins that helped children grow, broke down clots in the arteries of heart attack victims, and helped people with hemophilia clot their blood. Without having to rely on human blood to supply the clotting factors they needed, people with hemophilia were no longer at risk for HIV from blood transfusions.

But Boyer's and Cohen's studies also precipitated fears among scientists and the public that genetic engineering was an invasion of humanity into the realm of God. A *Time* magazine cover in the 1980s titled "Tinkering with Life" showed a DNA molecule surrounded by several white-coated scientists with hammers and rulers. At the head of the DNA was a fanged snake.

MERCK SCIENTISTS REALIZED THAT BOYER'S AND COHEN'S DISCOVERY could be used to make a hepatitis B vaccine without using human blood. They recruited a molecular biologist working at the University of California in San Francisco, William Rutter. Using Boyer's DNA-cutting enzyme, Rutter removed the surface protein gene from the virus and inserted it into one of Stanley Cohen's bacterial plasmids. When the bacteria reproduced, they made large quantities of hepatitis B surface protein. But Rutter and Merck found, much to their dismay, that the surface protein made by the bacteria didn't induce an immune response in animals. So they decided to try something else, soliciting the help of Ben Hall at the University of Washington. Hall used common baker's yeast instead of bacteria. Hilleman found that the hepatitis B surface protein made in yeast induced protective antibodies in chimps and, later, in people, so he used this system to make the next hepatitis B vaccine.

On July 23, 1986, the FDA licensed Merck's yeast-derived recombinant hepatitis B vaccine. The vaccine is still used today.

By the late 1980s, the hepatitis B vaccine had been used by less than 1 percent of the world's population. But between 1990 and 2000, hepatitis B vaccine usage increased to 30 percent. By 2021, more than 80 percent of the world's population had received the vaccine, and the impact has been dramatic. In Taiwan, hepatitis B vaccine has caused a 99 percent decrease in the incidence of liver cancer. In the United States, the incidence of hepatitis B virus infections in children and teenagers has decreased by 95 percent. Furthermore, because hepatitis B virus infects fewer people, the hepatitis B vaccine has dramatically increased the number of potential liver donors. "Hilleman's heroic role in controlling the hepatitis B virus scourge ranks as one of the most outstanding contributions to human health of the twentieth century or any century," recalls Thomas Starzl, a pioneer of liver transplantation. "From my parochial point of view, Maurice removed one of the most important obstacles to the field of organ transplantation."

Hilleman ranked the hepatitis B vaccine as his company's greatest single achievement: "We made the world's first hepatitis vaccine, the world's first anticancer vaccine, the world's first recombinant vaccine, and the world's first vaccine made from a single protein." If the worldwide use of hepatitis B vaccine continues, chronic infection with the virus will be virtually eliminated, and in thirty to forty years, so will consequent cirrhosis and liver cancer.

# CHAPTER 9

# Animalcules

*"You shall not crucify mankind upon a cross of gold."*

William Jennings Bryan

On Sunday morning, February 7, 1886, George Walker and George Harrison were strolling across the flat untouched savannah of South Africa. Walker was building a cottage for two brothers, the Stubens, and Harrison was building one for a widow, Petronella Oosthuizen. Idly kicking at the ground, Harrison stubbed his foot on an outcrop of rock. He picked it up, examined it carefully, pulled out his prospector's pickax, and struck off smaller pieces. Before coming to South Africa, Harrison had been a gold miner in Australia. Oosthuizen's nephew, George Overbay, remembered what happened next: "[Harrison] borrowed my aunt's frying pan in the kitchen, crushed the conglomerate to a coarse powder on an old ploughshare, and went to a nearby [water pump] where he panned the stuff. It showed a clear streak of gold."

On July 24, 1886, George Harrison wrote to the president of the South African Republic, Johannes Krüger, asking for a prospector's license. As news of Harrison's find spread, hundreds of miners rushed to the area, also hoping to get licenses. At nine o'clock on the morning of September 20, 1886, they got their answer. A government official stood beside his wagon and read Krüger's proclamation to the men: "I, Stephanus Johannes Paulus Krüger, advised by and with the consent of the executive council, proclaim [this district] as a public digging." Within a few months thousands of men had pitched tents in a town soon to be named Johannesburg. Three years later, Johannesburg was the most heavily populated city in Africa. By 1895, almost one hundred thousand people lived there. Of them, seventy-five thousand worked in the mines; all were poor black African men taken from their rural homes, separated from their wives and children.

Krüger, for whom the gold coin Krügerrand is named, was elected president of the South African Republic for the fourth and final time in 1898. Appalled by the greed of the mining companies, Krüger thought that residents should be lamenting, not celebrating, South Africa's find. "It will cause our land to be soaked in blood," he predicted. George Harrison, the discoverer of what was by the late 1930s the largest and richest gold mining area in the world, sold his license for £10. Several years later he was eaten by a lion.

British companies that owned the mines hired recruiters to deliver workers—derogatorily known as kaffirs—at a fixed price per head. Some became ill on the trip from their rural homes to the city. Others, crowded into small, poorly maintained barracks, suffered severe infections. Most suffered from malnutrition. If they survived, they worked for six to nine months before returning home. The constant turnover of miners meant the continuous introduction of new men into the mining compounds. Although these men suffered from dysentery and tuberculosis, no disease was more common, more severe, or more lethal than bacterial pneumonia. And every new miner was potentially susceptible.

In 1894, at a meeting of the Transvaal Medical Society of South Africa, doctors described an epidemic of a hundred cases of "purulent discharge from the nostrils and, in a large majority of cases, pneumonia." Fifteen of those men died. Five years later, doctors described a similar epidemic: "One batch of ninety-three emaciated Kaffirs arrived in the beginning of July and some of these were ailing; altogether of this batch, eight died." By the early 1900s, seven gold miners died of pneumonia every day. When doctors performed autopsies on men who had died and looked at sections of their lungs under the microscope, they found small round bacteria clustered in pairs. The name of the bacterium was *Streptococcus pneumoniae*, or pneumococcus. To prove that it killed the miners, researchers injected the bacteria into rabbits. Within days, all of the rabbits died.

Once they had identified the cause of this deadly pneumonia, researchers were ready to make a vaccine to prevent it.

THE FIRST VACCINE—EDWARD JENNER'S SMALLPOX VACCINE—PREVENTED a viral infection. Vaccines to prevent bacterial diseases like pneumococcal pneumonia lagged far behind, the first one appearing about a hundred years later; one of the reasons it took so much longer is that bacteria are much more complicated than viruses.

Viruses and bacteria are both made of proteins that evoke protective antibodies. But viruses don't contain many proteins; for example, measles virus contains ten proteins, and mumps virus contains nine. Bacteria are much larger; pneumococcus contains about two thousand proteins. Difficulties in determining which among these proteins evoked an immune response was among the reasons for the slower development of bacterial vaccines. Progress was also slowed by fraud.

IRONICALLY, RESEARCHERS KNEW ABOUT BACTERIA LONG BEFORE THEY knew about viruses. Martinus Beijerinck, investigating an infection of tobacco plants, was the first to figure out what viruses were and where they reproduced. But he never saw them. Not until the 1930s, with the invention of the electron microscope, did researchers finally see the viruses they were studying. Because bacteria were so much bigger than viruses, studies of bacteria had a three-hundred-year head start. In the late 1600s, Anton van Leeuwenhoek, a Dutch dry-goods dealer, produced the first microscope. While looking through his microscope at drops of rainwater or scrapings from his teeth, he noticed tiny creatures "moving in the most delightful manner." He called them animalcules—literally "little animals." We now know them in part as bacteria. It wasn't until the late 1800s that investigators found that bacteria weren't so delightful: some caused severe, often fatal illnesses.

The next breakthrough came when Robert Koch, a German bacteriologist, proved that specific bacteria cause specific diseases. In 1876 Koch set out to determine the cause of anthrax, a common and occasionally fatal lung disease in cattle but rarely in man. Living in the farmlands that housed his crude laboratory, Koch took pieces of spleens from cows that had died of anthrax and, using tiny wooden slivers, injected them into mice, all of which died. When he looked at the cow spleens through a microscope, they were teeming with bacteria. Koch reasoned that bacteria had killed the mice. Now he had to prove it. So he inoculated small pieces of spleens from infected cows into gelatinous fluid scooped from the center of an ox's eye, hoping that it would provide the nutrient substances necessary for the bacteria to grow. (Many early scientific studies sound like a witch's incantation from *Macbeth*.) During the next few weeks Koch watched the bacteria reproduce. He then injected his culture of anthrax bacteria into mice and found that again they all got sick; their lungs were loaded with anthrax bacteria. Koch had made an important observation. Until that time, scientists had believed that only bacteria taken from someone who was sick could make you sick. Koch proved that bacteria grown in his laboratory could also cause disease. Robert Koch was a father of the germ theory of disease.

During the next ten years Koch found that he could grow bacteria on nutrient media made from potatoes and gelatin. He placed his media in special flat glass dishes invented by a young researcher working in his laboratory, Julius Petri. Later, Koch discovered the bacteria that cause tuberculosis and cholera. By 1900, researchers had found twenty-one different bacteria that cause diseases. "As soon as the right method was found," said Koch, "discoveries came as easily as ripe apples from a tree."

Koch's observation that bacteria could be grown in the laboratory led to a series of important discoveries and three vaccines.

The first breakthrough occurred in the late 1800s, when two French researchers, Emile Roux and Alexandre Yersin, isolated the bacterium that causes diphtheria, then a common cause of death. Diphtheria causes a thick gray membrane to collect in the windpipe and breathing tubes, often suffocating its victims. In the United States alone, diphtheria infected two hundred thousand people every year, mostly teenagers, and killed fifteen thousand. Roux and Yersin, like Koch, found that they could reproduce the disease by injecting bacteria into experimental animals. But they also found that when they grew bacteria in liquid culture, the liquid alone caused severe and fatal disease; bacteria themselves weren't necessary. Apparently, diphtheria bacteria were making a toxin.

The second breakthrough resulted in the first Nobel Prize in medicine. Working in Marburg, Germany, Emil von Behring found that animals injected with diphtheria toxin made antibodies to the toxin, called antitoxin, and that antitoxin prevented disease. Scientists later extended Behring's discovery to make antitoxins to several bacteria. Behring's discovery was also the inspiration for the Iditarod dogsled race, which re-creates the life-saving emergency transport in 1925 of diphtheria antitoxin from Nenana to Nome, Alaska—a distance of 674 miles—during an outbreak of diphtheria. Although two children died during the outbreak, Behring's antisera saved the lives of many others.

Another French researcher, Gaston Ramon, made the third breakthrough in the late 1920s when he found that toxin that had been inactivated by formaldehyde could protect people against diphtheria. Now researchers no longer had to rely solely on antitoxins to fight bacterial infections. People injected with formaldehyde-treated toxin, known as toxoid, could be protected against diphtheria for the rest of their lives by making their own antibodies. This observation also led to vaccines against tetanus and, in part, whooping cough. Because of these three vaccines, the number of people killed every year in the United States by diphtheria decreased from fifteen thousand to five; by tetanus, from two hundred to fifteen; and by whooping cough, from eight thousand to ten.

Production of new bacterial vaccines exploded in the early 1900s. Pharmaceutical companies in the United States made vaccines by growing bacteria in pure culture, killing them with chemicals, and putting dead bacteria in a tablet. They called these vaccines bacterins. Bacterins were sold to prevent strep throat, acne, gonorrhea, skin infections, pneumonia, scarlet fever, meningitis, and intestinal and bladder infections. Bacterins were easily ingested, readily available, simple to make, and highly lucrative. There was only one problem: they didn't work. Nor did they have to. Pharmaceutical companies weren't required to prove that their products worked until the early 1960s. Change came later, but only when prompted by disaster.

IN 1954 CHEMISTS AT CHEMIE GRÜNENTHAL, A WEST GERMAN company, tried to make an antibiotic by heating a chemical called phthaloylisoglutamine. (Don't try to pronounce this word in your head.) The resultant drug didn't kill bacteria. So they tried something completely different: they tested animals to see whether the drug had an antitumor effect. Again, no luck. Finally, in a small test in people, researchers at Grünenthal found that the drug put patients into a natural, all-night sleep. On October 1, 1957, they advertised the drug as a sleeping pill and claimed that it was completely safe. They also claimed that pregnant women could use it to treat morning sickness, although they never specifically tested the drug for this use. They called the drug thalidomide. By 1960, hundreds of babies had been born with their hands and feet directly stuck to their bodies. Thalidomide damaged twenty-four thousand embryos; half died before birth. Today, about three thousand people live with birth defects caused by thalidomide.

The thalidomide disaster caused a reevaluation of the U. S. Food, Drug, and Cosmetic Act, passed in 1938. Congress amended it in 1962 to compel pharmaceutical companies to show that their products actually worked before selling them.

. . . .

THE FIRST PERSON TO TRY TO MAKE A vaccine TO PROTECT SOUTH African gold miners from pneumococcal pneumonia was Sir Almroth Wright. In February 1911, Julius Werhner, chairman of the Central Mining Investment Corporation in London, called upon Wright, a famous British researcher. A tough, opinionated man who actively opposed women's suffrage, Wright was the inspiration for the character of Sir Colenso Ridgeon in George Bernard Shaw's *The Doctor's Dilemma*. (The dilemma in Shaw's play is that Ridgeon, with enough antiserum to save one person from tuberculosis, must choose between a physician colleague and a talented artist. Smitten by the artist's wife, Ridgeon chooses the doctor, hoping that the artist will die as a consequence. Wright is said to have stormed out of an early performance of the play.) Werhner chose Wright to develop a vaccine against pneumococcus because he knew that several years earlier Wright had developed a successful vaccine against typhoid fever, caused by the bacterium *Salmonella typhi*. Wright had made his vaccine by growing *Salmonella* in pure culture and killing it with heat. Before his discovery, typhoid had been a common and fatal infection, especially among soldiers. During the Spanish-American War, about two hundred Americans died of their wounds, while typhoid killed two thousand. After Wright found that his vaccine worked, the British military gave it to all of its soldiers during the First World War. (The Second World War was the first in which more soldiers actually died in battle than of infection.)

Wright assumed that he could make a vaccine to prevent pneumococcal infection the same way that he had made his typhoid vaccine. So he took a strain of pneumococcus, grew it in culture, and killed it with a chemical. On October 4, 1911, Almroth Wright inoculated the first of fifty thousand South African gold miners with his vaccine. In January 1914 he published his results: "The comparative statistics that have been set forth above testify in every case to a reduction in the incidence rate and death rate of pneumonia in the inoculated." But Wright was wrong. One year later a statistician working for the South African Institute for Medical Research reanalyzed Wright's data and found that his vaccine didn't work at all. The reason for his failure would soon become evident.

In 1910, one year before Wright conducted his study, German researchers had found that there were two different types of pneumococci and that immunity to one didn't protect against infection from the other. By 1913, F. Spencer Lister, an English physician working in South Africa, had found four different types of pneumococci. More would follow. By the 1930s, researchers had identified thirty different types, and by the end of the Second World War, forty types. Today, researchers have identified at least ninety different types of pneumococci. Wright's attempts to protect South African gold miners failed because he didn't include enough different types of pneumococci in his vaccine. (Many of Almroth Wright's colleagues called him "Sir Almost Right.") Researchers had to find a way to prevent a disease that could be caused by many immunologically distinct types of the same bacterium.

Where Wright failed, Robert Austrian, a researcher at the University of Pennsylvania, succeeded. Austrian, who grew up in a three-story, high-ceilinged brick row house on Baltimore's Doctors' Row, was the son of Charles Robert Austrian, physician-in-chief at Sinai Hospital and an associate professor at Johns Hopkins. "I was scared to death to go into medicine," recalled Austrian. "My father was a hard act to follow." A trim, well-spoken man of German Jewish ancestry, Austrian finished his internship and residency at Hopkins in 1945 and decided to pursue a career in infectious diseases. "I was told that if you wanted to learn classical bacteriology, go to Harvard or [the] Rockefeller [Institute]. But if you wanted to see the future of bacteriology, go to New York University." At NYU Austrian met Colin MacLeod, the man who inspired his lifelong interest in pneumococcus. Austrian's interest in a pneumococcal vaccine would soon be interrupted by a medical product first developed in Germany.

IN 1908 A VIENNESE CHEMIST NAMED PAUL GELMO SYNTHESIZED A chemical that—because it was bright red—proved very useful in the

German dye industry. The dye also stained bacteria, which made them easier to see under a microscope. Twenty-five years later, a physician named Gerhard Domagk found that if he gave this dye to mice and rabbits, he could protect them against lethal doses of bacteria. Although the dye was effective in animals, Domagk hesitated to inject it into people. But in 1935, his young daughter became very ill with a streptococcal infection of her bloodstream, an often fatal disease. The crisis forced Domagk to act. In desperation, he gave her a dose of the dye, saving her life. The dye was called sulfanilamide, and it was the first antibiotic.

Austrian remembers the dawn of the antibiotic era: "Dr. Perrin Long [chairman of community medicine at Hopkins] was the first person to bring sulfa drugs back to the United States. The results of his treatment of several patients with streptococcal infections were so dramatic that some of his colleagues thought that he was lying. The possibilities seemed limitless." Sulfa drugs appeared to be the magic bullet in the fight against bacterial infections. At Hopkins, Perrin Long was the keeper of the gate, the man in charge of a very limited supply of sulfa drugs. He took his charge seriously, as demonstrated by one memorable incident. A co-worker remembered a late-night phone call in 1936. "I answered it, and a woman's voice asked for Dr. Long. He took the phone, and I heard him say, 'You can't fool me this time! I know you're not Eleanor Roosevelt,' and he hung up. Within seconds the phone rang again. This time he said meekly, 'Yes, Mrs. Roosevelt, this is Dr. Long.' The next day the newspapers announced that the president's son was ill. Later, they reported that the boy had been cured by sulfanilamide, supplied by [Dr.] Long."

By the early 1940s, researchers had found a method to mass-produce a different antibiotic, penicillin. Now, clinicians were confident that they could eliminate pneumococcal infections. "The drop in mortality was so dramatic," recalls Austrian, "that most people began to feel that this illness was no longer a common or serious one. And not only that: they no longer felt that it was necessary to identify pneumococci; the recognition of the organism declined. The opinion in the 1940s and 1950s was widely held that pneumococcal pneumonia had largely disappeared because of these new so-called wonder drugs." Austrian found, however, that while antibiotics saved lives, they didn't lessen the incidence of infection. "When I went from Hopkins to Kings County Hospital in Brooklyn, which was the third largest hospital in the United States, I was told that if I was really interested in pneumococcal pneumonia, I was probably in the wrong place because they saw so few cases every year." Austrian set up a laboratory to determine how many people with pneumonia were infected with pneumococcus. "The place was a madhouse," recalled Austrian. "There were four thousand beds. Beds were lined up in the corridors." He found that about four hundred people every year were admitted to Kings County with pneumococcal pneumonia. Austrian's colleagues remained unconvinced, believing that his findings were unique to Brooklyn. "So we got a grant from the National Institutes of Health to look at the incidence of pneumococcal pneumonia throughout the United States because, as a wag once said, 'Brooklyn is a city opposite the United States.' But it turned out there was just as much pneumococcal pneumonia coming into the city hospitals of Chicago, Los Angeles, and New Orleans as we had found in Brooklyn. It was very clear that the disease had not gone away. It was just a matter of seeing what you looked for."

After ten years of collecting data, Austrian found another surprise. He examined the death rates in three groups of patients with severe pneumococcal infections: those who had been treated with antibiotics, those treated with antiserum (made from the serum of horses injected with pneumococci), or those left untreated. Although antibiotics and antiserum clearly saved lives, they didn't work on the most severe infections. "The death rates among those who died in the first five days were essentially the same," said Austrian. "What this tells us is that if you are destined to die very early in illness, it won't make a difference what treatment we give you, because we don't understand even today [what's causing] early death. The only alternative then to protect those at high risk of early death is to prevent them from becoming ill." Austrian was talking about a pneumococcal vaccine.

. . . .

BETWEEN 1900 AND 1945, SCIENTISTS MADE SEVERAL IMPORTANT DIScoveries about pneumococcus. They found that the bacterium was surrounded by a capsule made of a complex sugar called polysaccharide, that the polysaccharide could be stripped from the bacteria, that polysaccharide injected into mice protected them against infection, and that people injected with different polysaccharides from various strains of pneumococcus developed antibodies against each of the different types. Colin MacLeod, Austrian's mentor at NYU, made the first successful pneumococcal vaccine by taking four different types of pneumococci and stripping off their polysaccharide coats. During the Second World War, he injected either his vaccine or placebo into seventeen thousand military recruits and found that following an epidemic of pneumococcal pneumonia, his vaccine worked. E. R. Squibb made MacLeod's vaccine in the late 1940s, using six different types of pneumococcus. Nobody bought it. Convinced that penicillin had eradicated pneumococcus, doctors weren't interested in a pneumococcal vaccine. So Squibb stopped making it.

By the early 1970s Austrian, now a professor at the University of Pennsylvania, decided to resurrect Colin MacLeod's pneumococcal vaccine. "I was told [by university administrators] that I could do anything that I wanted as long as I paid for it," he recalled. Austrian found that thirteen different types of pneumococcus accounted for a large percentage of the disease. So with support from the NIH, he made a vaccine containing polysaccharides from each of those types. Austrian and the NIH then convinced the pharmaceutical giant Eli Lilly to make thousands of doses of his vaccine. Confident that he could now protect people against pneumococcus, he called the medical directors of three of South Africa's largest gold mining companies and, on September 6, 1970, flew with his wife to Johannesburg to begin talks with them. That same day, members of the Popular Front for the Liberation of Palestine hijacked four planes bound for New York. "When we got off the plane we got an effusive welcome and we couldn't figure out why," recalled Austrian. "We were flying over North Africa when the planes were hijacked."

Two years later, Austrian began the first test of his vaccine at the East Rand Preparatory Mine in Boksburg, fifteen miles east of Johannesburg. First established in 1893, the mine was one of the oldest and deepest in the country. Although sixty years had passed since Almroth Wright had tested his vaccine, conditions in the mines weren't much better. Austrian decided to enroll men coming to work for the first time, reasoning that they were most likely to encounter types of pneumococci not found in their isolated rural communities. "It was an unforgettable experience," recalled Austrian, "to descend to the deepest workings of the mine two miles below the earth's surface and a mile below sea level." Rock temperatures reached 125 degrees.

Austrian wanted to inject some miners with his pneumococcal vaccine and some with placebo. But to convince gold mine executives to use his vaccine, he had to throw in another vaccine: one against meninogococcus, a different bacterium that caused rapid, overwhelming infection. Meningococcus scared mining company officials. Workers would be fine one minute and dead four hours later. "Death gave gold mining a bad name," recalled Austrian. "Money spent taking care of miners with pneumococcal pneumonia was just factored into the cost of the gold. But death from meningococcus was bad publicity. They let us do the [pneumococcal] study because of meningococcus." One third of the miners received pneumococcal vaccine, one third received meningococcal vaccine, and one third received placebo. (Like the pneumococcal vaccine, the meningococcal vaccine was made from the polysaccharide coat of the bacterium.) Although he wanted to see if his vaccine saved lives, Austrian also wanted to know what levels of antibodies in blood correlated with protection against disease. This meant that he needed to collect blood samples before and after vaccination. Austrian's intellectual interest angered officials from the mining company. "We wanted to study the serological responses to the vaccine," recalled Austrian. "The [company authorities] were willing to [allow miners to] give blood when they were acutely ill, but they objected to having

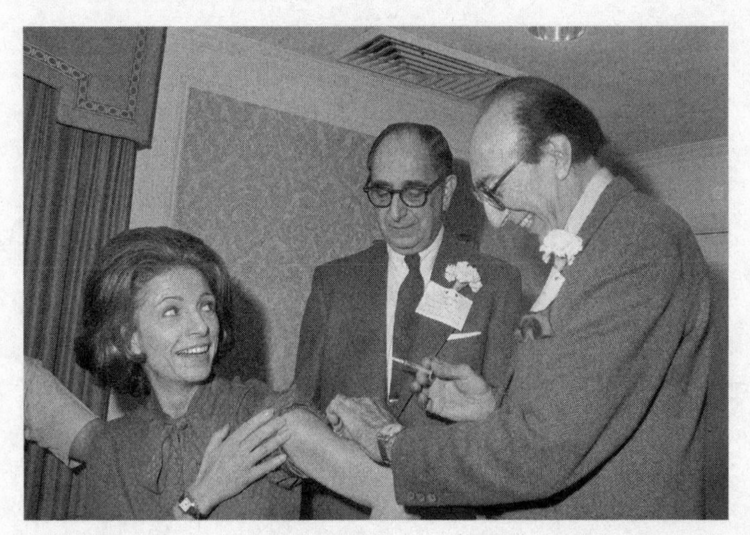

*Robert Austrian (center), winner of the 1978 Lasker Award, watches Dr. Michael DeBakey inoculate Mrs. William McCormick Blair, Vice President of the Lasker Foundation, with his pneumococcal vaccine, November 20, 1978.*

their blood drawn after they'd recovered. At one point I got a call from [the authorities] that they were going to call off the trial because it was interfering with the workforce of the mine. So I got on a plane on very short notice and went back to Johannesburg, where I met with the head of one of the mining companies. I pointed out to him that the vaccine had already saved him perhaps $100,000 in terms of medical expenses." To this, the mine owner slowly shook his head, realizing for the first time just how little Austrian knew about his business. "He looked at me, and he said that this [was] a $3 million annual operation. He wasn't interested in my problem. He had the coldest blue eyes I ever saw." The mining executive eventually decided to let the trial continue on the condition that Austrian abandon his interest in testing blood from the miners.

When the experiment was over, Austrian found that his pneumococcal vaccine worked, reducing the incidence of disease by 80 percent. Elated, he went back to the United States, certain that Eli Lilly would mass-produce the vaccine. But Lilly, which previously had been one of the largest vaccine makers in the United States, had decided to leave the vaccine business. With a vaccine in hand that could save thousands of lives, Austrian faced the very real possibility that no company would make it. In the end, only Maurice Hilleman considered Austrian's plea. "Dr. Hilleman on his own decided that Merck would make a vaccine," recalled Austrian. "If Maurice had said 'No,' it would have all gone down the drain. I don't know of anybody else in the vaccine business who was ready to step up if Maurice hadn't supported this."

In 1977 Maurice Hilleman and Merck made the first pneumococcal vaccine, designed to prevent infection with fourteen different types of pneumococci. In 1983 they made a second pneumococcal vaccine, containing even more different types. Austrian considers the pneumococcal vaccine to be one of the most unusual vaccines ever made. "This is probably the most complex vaccine that we have," says Austrian. "It's designed to protect against twenty-three different infections."

The CDC now recommends Robert Austrian's pneumococcal vaccine for certain high-risk groups, including people more than sixty-five years old. Unfortunately, many elderly adults in the United States don't get the pneumococcal vaccine, probably our most underutilized weapon in the fight against serious, and occasionally fatal, pneumonia. Worldwide, about two million people die of pneumococcal infections every year.

In addition to being among the first to make a vaccine against pneumococcus and later meningococcus, Hilleman and Merck were also among the first to make a vaccine against a bacterium called *Haemophilus influenzae* type b (Hib). Unlike pneumococcus, Hib, which causes severe meningitis, bloodstream infections, and pneumonia, preferentially kills young children. Unfortunately, researchers soon found that Austrian's idea of using bacterial polysaccharides to protect against diseases caused by bacteria like pneumococcus, meningococcus, and Hib didn't work in infants, who simply couldn't mount an immune response to bacterial polysaccharides. If Hilleman and others were going to protect young children against these bacteria, they were going to have to find a different way to do it.

In the late 1970s, John Robbins and Rachel Schneerson at the NIH and David Smith and Porter Anderson at Rochester University found that when they linked Hib polysaccharide to a protein, they could induce Hib antibodies in infants. The work inspired Merck and other companies to make a Hib vaccine. By the end of the twentieth century the incidence of Hib infections in young children had decreased by 99 percent.

The development of bacterial vaccines came just in time. The widespread use of a variety of different antibiotics has caused many bacteria, including pneumococcus, to become resistant to them. Unfortunately, pharmaceutical companies no longer devote much energy to making antibiotics. Vaccines may eventually stand alone as our last chance to fight bacterial infections.

Maurice Hilleman performed experiments critical to the development of the measles, mumps, rubella, hepatitis A, and hepatitis B vaccines, saving millions of lives every year. He was also the first to develop and mass-produce the pneumococcal and chickenpox vaccines and among the first to make the meningococcal and Hib vaccines. But from the 1990s into the twenty-first century, Maurice Hilleman, his vaccines, and his science would be at the center of a storm of controversy. Joe Lieberman, John Kerry, Dave Weldon, Dan Burton, Don Imus, Tim Russert, Robert Kennedy Jr., Doug Flutie, Anthony Edwards, and Cindy Crawford were among hundreds of politicians, sports figures, media personalities, and actors who stepped forward to oppose much of what Hilleman had accomplished.

# CHAPTER 10

## An Uncertain Future

*"To hear the allegation is to believe it. No motive for the perpetrator is necessary, no logic or rationale is required. Only a label is required. The label is the motive. The label is the evidence. The label is the logic."*

PHILIP ROTH, *THE HUMAN STAIN*

Vaccines face an uncertain future. On one hand, we can now make vaccines that are safer and better than ever before. In June 2006 vaccine makers licensed a vaccine to prevent infection with human papillomavirus (HPV), which causes cervical cancer, one of the most common cancers in the world. Cervical cancer kills about three hundred thousand women every year. After identifying which HPV protein evoked protective immunity, researchers took the gene that made the protein, put it into a plasmid, and put the plasmid inside common baker's yeast (the same strategy that was used to make Hilleman's hepatitis B vaccine). The yeast proceeded to make large quantities of the HPV protein. Then something fairly amazing happened: HPV proteins reassembled into a whole virus particle. Through an electron microscope, synthetic HPV was indistinguishable from natural HPV. The only difference between the two was that the synthetic virus didn't contain any viral DNA, so it couldn't possibly reproduce or cause disease. When this synthetic virus was given to thousands of women, it prevented HPV infection. The HPV vaccine was our second cancer vaccine (hepatitis B vaccine was the first).

The twenty-first century also witnessed the birth of two other vaccines, which protected against pneumococcal and meningococcal infections in children. The FDA licensed these two vaccines—both made by linking the complex sugar (polysaccharide) that surrounds bacteria to a harmless protein—in 2000 and 2005, respectively. With a vaccine to prevent pneumococcal infection, the incidence of pneumonia and bloodstream infections in children declined more than 90 percent. With the availability of a meningococcal vaccine, parents finally had a weapon against one of the most frightening diseases of children and teenagers. Every year in the United States meningococcus causes about three hundred cases of rapid, overwhelming bloodstream infections and meningitis. No infection causes greater panic in elementary schools, high schools, or colleges.

In February 2006 vaccine makers licensed a vaccine against rotavirus, one of the greatest killers of infants and young children in the world. Rotavirus attacks the small intestine, causing fever, vomiting, and diarrhea. Because the vomiting can be frequent, persistent, and severe, it's sometimes hard for children to recoup the fluids that they've lost. As a consequence, they can rapidly become dehydrated and die. In the United States alone, rotavirus causes one of every fifty infants to be hospitalized with severe dehydration. Worldwide, rotavirus kills six hundred thousand children every year—about two thousand every day. Scientists made the vaccine by capturing a strain of rotavirus that causes disease in cows but not people, purifying it, and combining it with human strains of rotavirus. Unlike the procedure with measles, mumps, rubella, and chickenpox vaccines, researchers first figured out which genes from human rotavirus made children sick and which genes caused a protective immune response. Now researchers no longer have to grow human viruses in animal cells and hope that they become weaker. They can simply figure out which genes make a virus dangerous and make sure that those genes aren't in the final vaccine. "Everything before was empirical and scientifically modest," says Adel Mahmoud, an Egyptian-born parasitologist and former president of the Merck Vaccine Division. "It was grow the bug, attenuate it, kill it, boil it, formalinize it, and it becomes a vaccine. The new vaccines launched at the beginning of this century are an order of magnitude above anything else that anybody would have dreamt of."

There's more good news. The gap in immunization rates between rich and poor countries is closing. In 1974, the WHO launched a program that increased immunization rates in the developing world from 5 percent to 40 percent. Perhaps the greatest success of the global immunization program has been the decline in the number of deaths resulting from measles—from eight million to less than two hundred thousand per year. Despite these successes, the WHO program has struggled to get vaccines to the children who need them. Of the one hundred thirty million children born every year in the world, between one million and two million still die of diseases preventable by vaccination. "We're perfectly willing to give away vaccines to the developing world," said one pharmaceutical company executive. "But we'd send vaccines to Africa and watch them fry on the tarmac." At the turn of the century, two people stepped forward with a plan to change that.

In 2000 Bill and Melinda Gates gave about $1 billion to create the Global Alliance for Vaccines and Immunizations (GAVI), with the caveat that the money could not be used until it was matched, dollar for dollar. UNICEF, the WHO, the World Bank, vaccine makers, and governments from ten countries met the Gateses' challenge. By December 2005, with more than $3 billion to spend, GAVI provided millions of doses of the hepatitis B, diphtheria-tetanus-pertussis, and Hib vaccines to the poorest countries. Immunization rates rose dramatically. "I'm a lot more optimistic than two decades back," says Mahmoud. "Investing in health is investing in the future. Not only does wealth make health, but health makes wealth."

We are now on the brink of successfully bringing vaccines to the developing world and closing a gap that has existed for centuries. We are, in many ways, at the dawn of a new age of vaccines. Unfortunately, during the past twenty years, forces designed to crush vaccines may have gained the upper hand.

. . . .

TOWARD THE END OF HIS LIFE MAURICE HILLEMAN FOUND HIMSELF in the middle of several controversies. The most enduring, mean-spirited, and sensational charge against him was that his vaccines, despite all of their success, caused autism.

Autism is an enigmatic disorder with a heartbreaking array of symptoms. Children with autism struggle to communicate with their parents, siblings, and classmates. They are often withdrawn; engage in repetitive, self-stimulating behaviors; eat poorly; and seem to live in a world of their own. Few things are more difficult for a parent then watching a child struggle to communicate. Autism became widely known in the early 1980s thanks to the television series *St. Elsewhere*, which depicted the lives and work of the staff of a fictional hospital, St. Eligius. The son of one of the main characters was severely autistic. During the final episode, the boy slowly turns over a snow globe containing a replica of the hospital, the years-long drama apparently a product of his wondrous imagination.

By the late 1960s Hilleman had decided to combine his measles, mumps, and rubella vaccines into a single shot, later known as MMR. He thought that getting one shot would be better than getting three. "It came out of a vision," said Hilleman, "a long-term dream that it might be possible one day to protect against these diseases in a single shot." Merck released Hilleman's MMR vaccine in the United States in 1971 and in the United Kingdom in 1988. Ten years later, a British researcher claimed that Hilleman's MMR vaccine had caused an epidemic of autism.

In February 1998 the prestigious Royal Free Hospital in London held a press conference. Journalists crowded into the meeting room, anxious to hear the results of a study soon to be published in the *Lancet*, a highly respected, widely read British medical journal. Sitting on a dais under the heat of television lights were five doctors, including the dean of the medical school. In the center of the group was the lead author of the study, Dr. Andrew Wakefield.

Wakefield was a compelling figure. Described by journalists as "tall, handsome, fluent, and charismatic [with] a sense of humor, a cultured British accent, and the body of a rugby player," Wakefield told the audience about eight British children in whom autism and intestinal problems had developed soon after they received the MMR vaccine. Wakefield reasoned that the measles vaccine contained in MMR, although injected into the arm, had damaged the intestinal lining, causing children to suffer abdominal pain and diarrhea. Because the intestinal surface no longer provided an adequate barrier, harmful proteins could enter the bloodstream and travel to the brain, where they caused autism. The study was full of holes. Wakefield didn't say what these proteins were; he hadn't identified them. He didn't say how measles virus given in the arm was destroying the intestine. He didn't say why measles vaccine would be more harmful when contained in MMR than when given alone. And, most important, Wakefield hadn't compared the incidence of autism in children who had received the MMR vaccine with those who hadn't. He only had a theory. But if he was right, autism now had a villain: Maurice Hilleman, the man who had decided to combine the three vaccines into one.

Parents of children with autism were intrigued by Wakefield's finding. Their children had been healthy, then received the MMR vaccine, then became autistic. Was this a coincidence? Or was the MMR vaccine really causing autism? Because about 90 percent of children in the United Kingdom had received the MMR vaccine soon after their first birthday, and because symptoms of autism typically appear when a child is between one and two years of age, it wasn't surprising that Wakefield found several children who had become autistic within one month of receiving the MMR vaccine. But anecdotal associations, which can be very powerful, can also be misleading. "We evolved to be skilled, pattern-seeking, causal-finding creatures," says Michael Shermer in *Why People Believe Weird Things*. "Those who are best at finding patterns—[for example] standing upwind of game animals is bad for the hunt; [or] cow manure is good for the crops—left behind the most offspring. We are their descendents. The problem in seeking and finding patterns is in knowing which ones are meaningful and which ones are not. Unfortunately our brains are not always good at determining the difference." One example of the power of anecdote can be found in the story told by a Philadelphia-area nurse practitioner. "A mother of a four-month-old brought her baby into the office for her shots," she recalls. "The baby was sitting on the mother's lap while I was drawing one of the vaccines into a syringe. I looked over and the baby started to seize. There was a history of seizures in the family, and the child went on to become epileptic. But imagine what the mother would have thought if I had given that vaccine five minutes earlier. She would have been convinced that the vaccine had caused her daughter's epilepsy. And all of the statistical data in the world showing her that it didn't wouldn't have convinced her otherwise."

To determine whether his anecdotal observations were correct, Wakefield needed to compare the incidence of autism in children who had received the MMR vaccine with the incidence in those who hadn't. But he never did this. In his *Lancet* paper, Wakefield understood that he had raised but not tested his hypothesis. "We did not prove an association between measles, mumps, and rubella vaccine and the syndrome described," he wrote. But during the press conference—when the lights were on him and he was speaking to anxious parents in the United Kingdom and the world—Wakefield abandoned all caution. He said that MMR should again be separated into three vaccines. "There is sufficient concern in my own mind for a case to be made for vaccines to be given individually at not less than one-year intervals," he said. "One more case of [autism] is too many. It's a moral issue for me and I can't support the continued use of these three vaccines given in combination until this issue has been resolved."

Hilleman had spent years trying to figure out how to combine the three viral vaccines into a single shot. He had determined the quantity of each vaccine virus that worked best and figured out how to stabilize the combination. He had compared immune responses in children who received his MMR vaccine with those who received the three vaccines separately. He had studied children to see how long their immunity lasted. Now Andrew Wakefield—in a single statement—had undone what it had taken Maurice Hilleman years to accomplish. Hilleman watched events in England unfold, powerless to do anything. "It saddened him," recalled a friend.

Wakefield was a media darling, combining the warm pathos of a caring physician ("Everything I know about autism, I know from listening to parents") with disdain for public health officials and pharmaceutical companies. "We are in the midst of an international epidemic [of autism]," he said. "Those responsible for investigating and dealing with this epidemic have failed. Among the reasons for this failure is the fact that they are faced with the prospect that they themselves may be responsible for the epidemic. Therefore, in their attempts to exonerate themselves, they are an impediment to progress. I believe that public health officials know that there is a problem; they are, however, willing to deny the problem and accept the loss of an unknown number of children on the basis that the success of public health policy—mandatory vaccination—by necessity involves sacrifice." To Wakefield, public health officials were knowingly sacrificing some children to autism so that others wouldn't suffer infectious diseases. At the time of Wakefield's pronouncements, British citizens had just seen health officials declare British beef to be safe during an epidemic of mad cow disease, only to watch these same officials ban beef from the marketplace months later. Certainly, if officials were willing to cover up their failure to find the source of mad cow disease, they would be willing to cover up their failure to recognize that vaccines caused autism.

The day after the Wakefield press conference, with the imprimatur of a leading hospital, a respected medical journal, and a call by the lead investigator for immediate government action, the British media exploded. Headlines in *The Guardian* and *Daily Mail* read "Alert over Child Jabs" and "Ban Three-in-One Jab, Urge Doctors." Tony Blair, Britain's prime minister, fueled the controversy by refusing to say whether his son, Leo, had received the vaccine, claiming it to be a private family matter.

Parents in England and Ireland responded by refusing to give their children the MMR vaccine. In the months following Wakefield's announcement, one hundred thousand parents chose not to vaccinate their children. As a consequence, the incidence of measles in England and Ireland skyrocketed. "After the study first came out we were struggling just to get parents to immunize their children," said Michele Hamilton-Ayers, a pediatrician in Cheltenham, England. "Things got terribly bad." Measles, a disease that vaccines had easily controlled, was back. Within months, one small children's hospital just outside of Dublin admitted a hundred children with measles, three of whom died. One was a fourteen-month-old girl, Naomi. "I couldn't believe that this could happen," said her mother, Marie. "We used to hear about measles, but I never thought that it could be this bad." Suffering from high fever and with difficulty breathing, Naomi had been admitted to the hospital's intensive care ward with measles pneumonia. After three weeks she went home, but she wasn't better. "She started twitching with her eye," said Marie. "I couldn't believe that there was something wrong with her again." Having infected Naomi's lungs, measles virus now infected her brain. Naomi went back to the hospital. "By the time we got up to the ward she was dead. They were taking all the tubes out of her. When she first got sick, the nurse said that it was only the measles. Only?"

The allegation that MMR vaccine might be causing autism spread quickly to the United States. Several politicians took up the cause. On April 12, 2000, Dan Burton, chairman of the House Committee on Government Reform, brought together scientists, public health officials, doctors, and parents to get at the truth, to sort out what was behind this epidemic of autism. He began the congressional investigation with a statement: "I'm very proud of that picture," he said, pointing to a photograph, projected onto a large screen at the front of the room, of his granddaughter Alexandra and his grandson Christian. "The one on the left is my granddaughter. She almost died after receiving the hepatitis B shot. Within a short period of time she quit breathing and they had to rush her to the hospital. My grandson—who you see there with his head on her shoulder—according to the doctors was going to be about six feet ten. We anticipated having him support the family by being an NBA star. But unfortunately, after receiving nine shots on one day, the MMR and the DTaP and the hepatitis B, within a very short period of time he quit speaking, ran around banging his head against the wall, screaming and hollering and waving his hands, and became totally a different child. And we found out that he was autistic. He was born healthy. He was beautiful and tall. He was outgoing and talkative. He enjoyed company and going places. Then he had those shots and our lives changed and his life changed." Burton stopped, trying to compose himself. "I don't want to read all of the things that happened to Christian," said Burton, "because I don't believe that I could make it through it. But I can't believe that this is just a coincidence. That the shot is given and that within a very short period of time instead of being the normal child that we played with and talked to, he was running around banging his head and flailing his arms. And when people tell me that that's just a genetic problem I'm telling you that they're just nuts. That's not the way it was."

Burton, with the help of antivaccine activists, had called several people to testify in front of his committee. Andrew Wakefield talked about children who had recently been vaccinated with MMR vaccine and had become autistic. John O'Leary, a molecular biologist from Ireland, showed pictures of measles virus proteins in the intestines of Wakefield's autistic children. O'Leary failed to mention that other investigators evaluating those same samples couldn't find what he had found, some questioning whether O'Leary had made up his data. Burton watched slide after slide as Wakefield and O'Leary described polymerase chain reactions, fusion proteins, hybridization analyses, and the importance of follicular dendritic cells. Clearly in over his head, Burton later remarked that perhaps these results could be explained in simpler terms.

Not all who testified were willing to line up behind Burton. Brent Taylor, an epidemiologist also working at the Royal Free Hospital in London (where Wakefield had performed his study), testified that the rates of autism in children who did or did not receive MMR vaccine were the same. This wasn't how Burton wanted the meeting to go. Reading a statement prepared for him by antivaccine groups as a rebuttal to Taylor's arguments, Burton asked whether Taylor had arbitrarily excluded some cases of autism. When Taylor responded that he had included every case of autism that he had identified, Burton was stuck. He hadn't read either Wakefield's or Taylor's papers.

Henry Waxman, the ranking Democrat on the committee, questioned the venue for such a discussion. "I'm troubled by this meeting," said Waxman. "This hearing was called and structured to establish a point of view. And that's the point of view of the chairman [Burton]." Waxman suggested that Burton had stacked the deck by excluding groups that had asked to testify at the hearing—among them the American Medical Association, the American Public Health Association, the Infectious Diseases Society of America, the American Nurses' Association, Britain's Medical Research Council, the WHO, and Louis Sullivan, the former head of the Department of Health and Human Services. "I think hearings like this have a real danger," said Waxman. Later in the hearing, Waxman again questioned the validity of Congress as the place to determine scientific truths. He asked that scientific studies be performed and evaluated by scientists and that congressmen defer to the scientific process. "Let's let the scientists find where the truth may be," he said. To Waxman, having politicians judge the science of vaccines in Congress made about as much sense as having scientists legislate voters' rights in their laboratories.

Burton wasn't the only politician who supported the notion that MMR caused autism. Congressman Dave Weldon wrote a letter to Louis Cooper, then president of the American Academy of Pediatrics: "I am compelled to urge the Academy to recommend to pediatricians that they inform parents of their option of separating the MMR for their children. I am also contacting public health officials to urge that they examine their policies as well."

The American media loved the story. The *New York Times*, *CNN*, *USA Today*, the *Washington Post*, and almost every major newspaper, magazine, and radio and television station in the United States elevated Wakefield's hypothesis to fact: MMR caused autism. On November 12, 2000, the television program *60 Minutes* produced a segment titled "The MMR Vaccine." Ed Bradley was the correspondent. The segment began with Bradley interviewing Dave and Mary Wildman, the parents of an autistic son, from Evans City, Pennsylvania. Bradley said that the boy "appeared perfectly normal until just after his first birthday, when he received the MMR

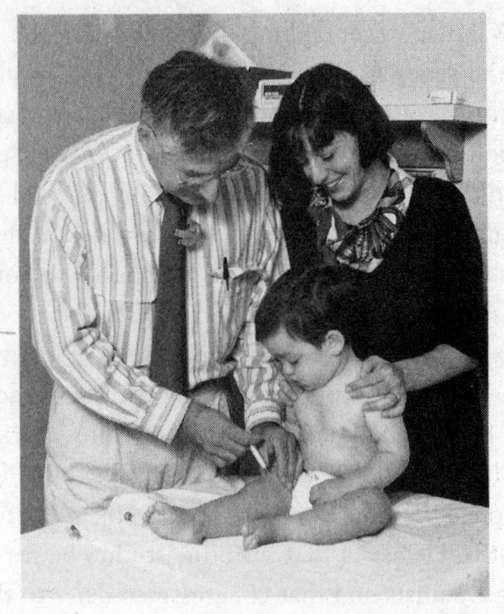

*Dr. Richard Buchta (left) inoculates Jeryl Hilleman's son Colin with the MMR vaccine, 1991.*

vaccine. Within a few weeks, according to his parents, things began to change." "He started to not look at me anymore when I would call his name," said Mary. "And do you know why?" asked Bradley. "Because of the MMR vaccine," she said, tears streaming down her face. "I should never have had him have that vaccine." Bradley also interviewed Andrew Wakefield during the segment. "My concern comes initially from the story the parents tell," said Wakefield. "They have a normally developing child who upon receipt of the MMR vaccine develops a complex syndrome of behavioral and developmental regression, loss of speech, loss of language, loss of acquired skills, loss of socialization with siblings or peers." "Do you have children?" asked Bradley. "I have four children," Wakefield replied. "Knowing what you know now," said Bradley, "would you give them the MMR vaccine?" Wakefield leaned toward the camera: calm, certain. "No, I wouldn't," he said. "I would most certainly vaccinate them. I would give them [separate shots of] measles, mumps, and rubella vaccines." After watching the program, Marie Lynch, a thirty-two-year-old event planner from Chicago, said, "I did permit my two-year-old daughter, Tess, to be vaccinated with MMR. However, I crossed my fingers and prayed for days."

In response to the growing controversy, epidemiologists, researchers, and public health officials in the United States, Denmark, the United Kingdom, Finland, and other countries sorted through medical records, trying to determine whether children who had received the MMR vaccine were at greater risk of autism than were those who hadn't. During the next few years, fourteen separate groups of investigators evaluated the records of more than six hundred thousand children. The results were clear, consistent, and reproducible: the incidence of autism was the same in both groups. MMR vaccine didn't cause autism. Parents choosing not to immunize their children weren't reducing their risk of autism; they were only increasing their risk of catching potentially fatal infections. Andrew Wakefield's conjecture hadn't stood up to further scrutiny.

In February 2004 Brian Deer, an investigative reporter in London, found that Andrew Wakefield wasn't exactly what he appeared to be. Writing for the *Sunday Times* of London, Deer found that Wakefield's *Lancet* paper contained several errors. In its acknowledgments the paper stated that "this study was supported by the Special Trustees of Royal Free Hampstead NHS Trust and the Children's Medical Charity." But Wakefield had omitted the study's largest supporter. Two years before his paper was published, Wakefield had signed a contract worth £55,000 with a personal injury lawyer named Richard Barr. Five of the eight autistic children in Wakefield's study were Barr's clients. Wakefield knew that the parents of the children in his study had a financial interest in finding a link between the MMR vaccine and autism. If Wakefield was successful in establishing that link, these parents could successfully sue for compensation. Wakefield chose not to disclose his financial ties to Barr, either to the editor of the *Lancet* or to his fellow investigators.

The acknowledgments weren't the only part of Wakefield's paper that were misleading. Wakefield claimed that he had first encountered children during their routine admissions to the hospital, when in fact he had been alerted to their existence by Barr. He had then laundered the parents' claims about their children's histories into clinical findings to fit a medical publication, grossly misleading the public. In the wake of the *Lancet* study, seven hundred families banded together in London to sue pharmaceutical companies on behalf of their autistic children; many of these families were clients of Richard Barr.

Finally, Wakefield claimed that "investigations were approved by the Ethical Practices Committee." But the ethics committee had never approved the study. Further, Wakefield later reassured the committee that all of the invasive tests performed on these children—including blood collections, spinal taps, colonoscopies, and intestinal biopsies—would have been carried out even if these children weren't being studied. This was clearly untrue. The fact that young children with autism were being subjected to biopsies and spinal taps for the purpose of generating evidence for a lawsuit caused many to wonder exactly who was looking out for their well-being.

Looking for a response to Deer's allegations, reporters cornered Wakefield during a press conference. Wakefield admitted that "four, perhaps five" of the children in the *Lancet* study were clients of Richard Barr. "Was it four or five?" they asked. "Let's make it five," he said. "Were they litigants?" "Yes," replied Wakefield. "Were you being paid to help them build their case?" Again Wakefield said yes. "Did you tell your colleagues that these children were part of the study?" "I don't recall," said Wakefield. "Did you tell the *Lancet* about these conflicts prior to the publication?" Wakefield said that he hadn't. "Why not?" "I believe that this paper was conducted in good faith," said Wakefield. "It reported the findings. There was no conflict of interest." "Do you have any reasons now to change your opinion?" "No," said Wakefield.

Sir Liam Donaldson, the British government's chief medical officer, saw Wakefield's report for what it was: bad science. By failing to examine the incidence of autism in vaccinated *and* unvaccinated groups, Wakefield hadn't studied anything; he'd merely advanced a hypothesis based on a few children. Donaldson felt that Wakefield's paper should never have been published, not because of conflicts of interest but because it failed to shed any light on the cause or causes of autism. He recognized that Wakefield's pronouncements came at great cost. Speaking on BBC's *Today* program, Donaldson said, "If the paper had never been published, then we wouldn't have caused a completely false loss of confidence in a vaccine that has saved millions of children's lives around the world."

When the *Sunday Times* revealed that a personal injury lawyer had supported Wakefield's study, Richard Horton, the editor of the *Lancet*, was shocked. "There were fatal conflicts of interest in this paper," he said. "If we had known [about them], it would have been rejected. As the father of a three-year-old who has had the MMR, I regret the adverse impact this paper has had." Simon Murch, one of the co-authors of Wakefield's paper, was also caught off guard: "We never knew anything about the £55,000—he had his own separate research fund. All of us were surprised. We were pretty angry." Six years after Wakefield's publication, in September 2004, ten of the thirteen original authors withdrew their support of Wakefield. In a strongly worded letter to the *Lancet* they said, "We wish to make it clear that in this paper no causal link was established between the vaccine and autism, as the data were insufficient. However, the possibility of such a link was raised. Consequent events have had major implications for public health. In view of this, we consider now is the appropriate time that we should together formally retract the interpretation placed on these findings in the paper."

The Royal Free Hospital fired Andrew Wakefield, and the General Medical Council in England filed eleven charges against him for misconduct. Wakefield sought refuge in the United States, first working in Florida and later in Texas, constantly lecturing an anxious public about the dangers of vaccines. To many, Wakefield remains a hero, a tragic figure bravely standing up against a medical establishment determined to crush him.

During the *60 Minutes* interview with Ed Bradley, Wakefield said, "I would have enormous regrets if [my theories] were wrong and there were complications or fatalities from measles." Wakefield was right in predicting that parents would soon watch children suffer and die of measles, but he overestimated his capacity for regret. Although study after study showed that MMR never caused autism, Wakefield remains unrepentant, wedded to a hypothesis that he considers nonfalsifiable. Still seeing himself as a champion of children, a man devoted to keeping them safe from vaccines foisted on the public by greedy pharmaceutical companies and inept public health officials, he asks, "Should we stop, should we go away, should we stop publishing because it is inconvenient? I've lost my job. I will never practice medicine in [England] again. There is no upside to this. But if you come to me and say, 'This has happened to my child,' what's my job? What did I sign up to do when I went into medicine? I'm here to address the concerns of the patient. There's a high price to pay for that. But I'm prepared to pay it."

Andrew Wakefield remains a man committed to saving children from the "harm" of the MMR vaccine, even though no harm exists. And because it's hard to unring the bell, some parents in the United States, England, and the world still refuse to give the MMR vaccine to their children, fearing that it causes autism.

WHEN SCIENTIFIC EVIDENCE CONVINCINGLY REFUTED WAKEFIELD'S notion that MMR caused autism, antivaccine activists in the United States didn't stop. They shifted their vaccines-cause-harm hypothesis to thimerosal, a mercury-based preservative contained in some vaccines. Thimerosal, they said, was causing autism. Again, Hilleman found himself in the middle of the fray.

The controversy started innocently enough. In 1997 Congress passed the FDA Modernization Act. The bill, which received little attention from the media at the time, required health officials "to compile a list of drugs and foods that contain intentionally introduced mercury compounds and [to] provide a quantitative analysis" of those compounds. In response, health officials started adding up the amount of mercury in a variety of medical products, including vaccines.

At the time of the bill, the most common preservative used in vaccines was thimerosal. Because of earlier tragedies, pharmaceutical companies had been adding thimerosal to vaccines since the 1930s. Before adding thimerosal, between 1900 and 1930, companies packaged vaccines almost exclusively in multidose vials. A typical vial contained ten doses. Because a large percentage of the cost of vaccines is determined by the cost of the packaging—sterile glass vials, rubber stoppers, metal tops, and labels—as well as the labor required to fill each vial, multidose vials allowed vaccines to be made less expensively. Doctors kept these vials in their office refrigerators. When they gave vaccines to children, they would insert a needle through the rubber stopper, pull up the liquid into a syringe, and inject it into the arms of their patients. Unfortunately, by constantly violating the rubber stopper with a needle, doctors and nurses would inadvertently contaminate the vial with bacteria. Children given the eighth, ninth, or tenth dose from a vial were at greater risk of bacterial infection than those given the first dose. Bacteria that contaminated vaccines caused abscesses at the site of injection, as well as severe and occasionally fatal infections. In the first two decades of the twentieth century at least sixty children were known to have died of bacterial infections caused by multidose vaccines that didn't contain preservatives.

Early in the twentieth century, scientists found that small quantities of mercury prevented bacteria from growing. Although it was also known that large quantities of environmental mercury—known as methyl mercury—could cause permanent brain damage, small quantities appeared to be harmless. As an added precaution, pharmaceutical companies chose a type of mercury not found in the environment—ethyl mercury—because it was eliminated from the body much more quickly than environmental mercury but still killed bacteria. (Some people might remember the mercury-containing orange-colored antiseptic Mercurochrome, used to prevent bacterial contamination of cuts and scrapes in the 1950s, 1960s, and 1970s.) Although ethyl mercury and methyl mercury sound similar, they're quite different. An analogy can be made between ethyl alcohol and methyl alcohol. Ethyl alcohol, contained in wine or beer, can cause headaches and hangovers. Methyl alcohol, otherwise known as wood alcohol, causes blindness. By the late 1930s, ethyl mercury had been added to several vaccines, and infections caused by contaminating bacteria virtually disappeared.

During the next seventy years, pharmaceutical companies made many new vaccines, packaged them in multidose vials, and added thimerosal as a preservative. But as children received more and more vaccines, the quantity of mercury they received also increased. When the FDA determined the amount of mercury contained in vaccines in the late 1990s, it found that infants could be injected with as much as 187.5 micrograms. (A gram is approximately the weight of one-fifth of a teaspoon of salt. A microgram is one-millionth of a gram.) Officials consulted federal safety guidelines for ethyl mercury and found that there were none; guidelines existed only for methyl mercury. So they decided to use the guidelines established for methyl mercury to determine the safety of ethyl mercury. Public health officials consulted safety guidelines from three groups: the FDA, the Environmental Protection Agency (EPA), and the Agency for Toxic Substances Disease Registry (ATSDR). They found that the amount of mercury that children received in vaccines was two times greater than safety levels recommended by the EPA but didn't exceed those recommended by the FDA or ATSDR. The American Academy of Pediatrics (AAP) and the federal Public Health Service (PHS) saw the intentional administration of mercury to children at levels that exceeded EPA safety guidelines as a public relations nightmare. In October 1999 the two organizations issued a statement that they hoped would "maintain the public's trust in immunization": they asked that thimerosal be removed from vaccines as quickly as possible. The advisory stated that thimerosal wasn't being removed because it was known to be harmful but because it would make "safe vaccines safer." Critics wondered how removing an ingredient that hadn't been shown to be harmful could make a vaccine safer. Subsequent events would show that this attempt to reassure the public will probably remain forever as an example of how not to communicate theoretical risks.

Pharmaceutical companies complied with the AAP-PHS directive to remove thimerosal from vaccines. With the exception of the influenza vaccine, they eliminated it by packaging vaccines in single-dose vials. Although health officials tried to reassure the public that thimerosal hadn't been shown to be harmful, parents wondered why pharmaceutical companies would act so precipitously to remove it if there wasn't a problem. Again, no one jumped on this series of events more quickly, more passionately, or more effectively than a small but vocal group of parents of children with autism.

In the wake of the removal of thimerosal from vaccines, several powerful forces came together. Parents of children with autism saw the mercury debate as a possible solution to their problems. If mercury caused autism, then using chemicals that removed mercury from their children might help. Personal injury lawyers saw the controversy as a large and tempting pool of money. If mercury caused autism and if pharmaceutical companies knew that they had exceeded federal safety guidelines, companies would be liable for damages. Given that tens of thousands of children were diagnosed with autism every year, the money to be made from settlements and awards seemed limitless. The media saw the issue as a great man-bites-dog story: vaccines, claimed for years to be a life-saving product, were actually causing harm. And politicians saw the vaccines-cause-autism controversy as a way to show sympathy for grieving parents while drawing the light of television cameras. All they had to do was to ban mercury-containing vaccines from their states—politically, not a very heavy load to lift.

In the summer of 2005, all of these forces erupted on the American scene.

Robert F. Kennedy Jr., supported by plaintiff lawyers, wrote an exposé for *Rolling Stone* magazine titled "Deadly Immunity." In his article, Kennedy painted a picture of greedy pharmaceutical companies, sheep-like doctors, and public health officials trying to cover up a problem that had occurred on their watch—and was now spiraling wildly out of control.

Arnold Schwarzenegger of California became the first governor to prohibit vaccines containing thimerosal from his state. Others soon followed his lead. Because the supply of thimerosal-free influenza vaccine was limited, Schwarzenegger had essentially prohibited many of California's children from receiving a vaccine that prevented influenza—a disease that every year in the United States still causes the hospitalization of one hundred thousand children and the deaths of a hundred.

Finally, the advocacy group Safe Minds commissioned a reporter, David Kirby, to write a book about the thimerosal-autism controversy. Kirby had never written a story about health, science, medicine, or vaccines before. But with funding from a wealthy California financier, the book, *Evidence of Harm*, became one of the best-selling health books in the United States. Don Imus interviewed Kirby several times on his national radio program. Tim Russert interviewed him on *Meet the Press*. Indeed every major television and radio outlet trumpeted the *Evidence of Harm* story: a story of intrigue, secret meetings by the CDC, and public health officials asleep at the switch. Everyone, it seemed, was in the pocket of pharmaceutical companies. For months after the publication of *Evidence of Harm*, the thimerosal-autism story dominated the news. The media portrayed the story as little guy (parents) versus big guy (pharmaceutical companies)—a story they knew most Americans would love.

The growing fear that mercury in vaccines caused autism soon led to tragedy. On April 3, 2005, a five-year-old autistic boy, Tariq Nadama, visited the office of Dr. Roy Eugene Kerry in Portersville, Pennsylvania. Kerry rolled up Tariq's sleeve, inserted a needle into his vein, and injected him with ethylenediaminetetraacetic acid (EDTA), a chemical that binds to mercury and helps to remove it from the body. EDTA therapy has never been shown to improve symptoms of autism, is not approved by the FDA for this purpose, and can be quite dangerous. At the time, about ten thousand children received mercury-binding therapy every year, spurred largely by the precipitous removal of thimerosal from vaccines. There remains no evidence that this therapy helps. Five minutes after the injection Tariq had a heart attack and died.

Supported by the media attention that they had helped to generate, personal injury lawyers filed 350 claims in federal and state courts seeking billions of dollars in compensation. Maurice Hilleman soon found himself in the middle of the fray. A Washington-based personal injury lawyer named James A. Moody leaked a memo to Myron Levin, a medical writer for the *Los Angeles Times*. In the early 1990s Hilleman had written the memo to Gordon Douglas, then the head of the Merck Vaccine Division. Moody claimed that an unnamed whistle-blower, stricken by conscience, had given the memo to him. (Actually, the memo had been given to plaintiffs' lawyers as part of the normal discovery process prior to trial.) In the memo Hilleman stated that "the mercury load [in vaccines] appears rather large." Personal injury lawyers used Hilleman's memo as the smoking gun that proved that companies were aware of the unacceptably high levels of mercury in vaccines well before the FDA Modernization Act. But Hilleman also stated in his memo that "the key issue is whether thimerosal, in the amount given with the vaccine, does or does not constitute a safety hazard. However, perception of hazard may be equally important." Hilleman, like the AAP and PHS, was concerned about the media and advocacy groups. But because Hilleman also knew that mercury was present in the environment, he wasn't sure whether the quantity of mercury contained in vaccines was an important addition. "It appears essentially impossible based on current information," he wrote, "to ascertain whether thimerosal in vaccines constitutes or does not constitute a significant addition to the normal daily input of mercury from diverse sources."

After writing his memo, Hilleman reviewed more recent studies and concluded that the quantity of mercury in vaccines wasn't harmful, a belief supported by several facts.

Mercury, part of the earth's crust, is released into the environment in the form of inorganic mercury by volcanic eruptions, burning coal, and water erosion of rocks. It is then converted from inorganic to organic mercury—specifically, methyl mercury—by bacteria in the soil. Methyl mercury then enters the water supply and eventually the food chain.

Because methyl mercury is everywhere, it's unavoidable. Mercury is in water, infant formula, and breast milk. A breast-fed child ingests about 360 micrograms of methyl mercury in the first six months of life, twice the quantity of mercury contained in vaccines before its removal. And the mercury in breast milk is methyl mercury, excreted much more slowly from the body—and therefore much more likely to accumulate—than the ethyl mercury contained in vaccines.

Because babies encounter far more mercury in their environment than they do in vaccines, Hilleman believed that mercury in vaccines was never harmful, but he was savvy enough to know that it might be perceived as harmful. Nevertheless, Hilleman later regretted writing the memo. "I didn't think that it would get into the hands of people who couldn't think at all. Myron Levin [from the *Los Angeles Times*] never talked to me. And there was no whistle-blower at Merck. They misinterpreted what I was saying. My concern was that the public wouldn't know the difference between ethyl mercury and methyl mercury, because public perception is not often informed by science. But then the personal injury lawyers got a hold of this."

Later, five studies performed on three continents clearly showed that the incidence of autism was the same in children who received vaccines that did or did not contain thimerosal. The Institute of Medicine, an independent organization within the National Academy of Sciences, reviewed these studies and concluded that thimerosal didn't cause autism. Perhaps the best study, published in July 2006, took advantage of a natural experiment that occurred in Canada between 1987 and 1998, when the quantity of thimerosal in vaccines varied. Between 1987 and 1991, vaccinated babies in Montreal received 125 micrograms of thimerosal; between 1992 and 1995, they received 225 micrograms; and after 1996 they received 0 micrograms. If thimerosal caused autism, the incidence of autism should have been much higher in children born between 1992 and 1995 than in those born after 1995. In fact, the opposite was true. The incidence of autism was higher in babies born after 1995 than in those born before 1995. Similarly, studies in Denmark, a country that had abandoned thimerosal as a preservative in 1991, found an increase in autism several years later. These increases in autism rates were most likely due to a broadening of the definition of the disease to include autistic spectrum disorder, Asperger's syndrome, and pervasive developmental disorder.

Although thimerosal at the level contained in vaccines doesn't cause autism, the financial incentives of those interested in keeping this controversy alive make it unlikely that it will die down anytime soon.

. . . .

In addition to antivaccine activists, personal injury lawyers, and an occasionally irresponsible media, other obstacles stand in the way of vaccines. For example, we are choosing not to afford them.

In the early 1950s, there were four vaccines: diphtheria-tetanus-pertussis (DTP) and smallpox. Together, these vaccines cost less than $2, and people paid for them out of their own pockets. About 40 percent of children in the United States were immunized—a rate similar to that in many developing countries today.

In the 1970s, there were seven vaccines: DTP, MMR, and polio. The price for all of these vaccines was less than $50. People still paid for the vaccines themselves, and immunization rates rose to 70 percent.

By the mid-1990s, there were ten vaccines: DTaP (which included the new pertussis vaccine), MMR, polio, Hib, hepatitis B, and chickenpox. The price for these vaccines had risen to hundreds of dollars, and many parents couldn't afford them. With more vaccines at greater cost, the large private market for vaccines was dwarfed by another payer: the federal government. In the early 1990s, Bill Clinton found it unconscionable that uninsured or underinsured children in the United States weren't getting the vaccines that they needed. So he created the federal Vaccines for Children (VFC) program. With federal funds in place, immunization rates rose from 70 to 90 percent, and several diseases were completely or virtually eliminated.

By 2006, the CDC had placed sixteen vaccines on the routine schedule, adding rotavirus, meningococcus, influenza, papillomavirus, pneumococcus, and hepatitis A. The cost for these vaccines was more than $1,000. The federal government and insurance companies were now spending about $3 billion a year on vaccines. (Although $3 billion sounds like a lot of money, it's only about one-tenth of 1 percent of the almost $3 trillion that the United States spends every year on health care.) Unfortunately, at a time when vaccines required the most support, the federal government became an unreliable and inconsistent buyer, cutting back on a program that gave money directly to states to buy vaccines. Vaccine funding was at a crisis. As a consequence, many states that had previously purchased all vaccines for their children couldn't afford them. Forced to choose, some states didn't provide the pneumococcal vaccine, and others didn't provide the meningococcal or papillomavirus vaccines. States were forced to decide which infectious diseases they would prevent and which they would allow to continue to cause suffering and death.

Our willingness to pay for the treatment of people who are sick but not to prevent disease in those who aren't is rooted in the myth of invulnerability.

WE NEVER BELIEVE THAT A DISEASE IS GOING TO HAPPEN TO US—UNTIL it happens to us.

In the 2002 movie *John Q*, Denzel Washington plays a father, John Quincy Archibald, whose son needs a heart transplant. If the boy doesn't get the transplant, he'll die. John tries patiently, then desperately, to get his insurance company to pay for the transplant. But it refuses. Because John can't afford the transplant and because the hospital surgeons are unwilling to perform the operation for free, the boy is essentially sentenced to death. Like any parent, John cannot stand by and watch his son die. So he takes over the hospital at gunpoint. Anyone watching this movie can identify with his anguish. But what if there were a vaccine that prevented heart damage? Would John have been as passionate, committed, and ultimately crazed years earlier if his son had been denied the vaccine, not knowing for sure that he might benefit?

That's the problem with vaccines. When they work, absolutely nothing happens. Nothing. Parents go on with their lives, not once thinking that their child was saved from meningitis caused by Hib or from liver cancer caused by hepatitis B or from fatal pneumonia caused by pneumococcus or from paralysis caused by polio. We live in a state of blissful denial. But somebody was getting those diseases. Before pharmaceutical companies made the Hib vaccine in the early 1990s, every year about ten thousand children were stricken with meningitis, leaving many blind, deaf, and developmentally disabled. Today, fewer than fifty children every year suffer this disease. But who are the thousands of children who aren't getting Hib today? What are their names? We don't know. And that's what makes vaccines—or any prevention—so much less compelling than treatment. We spend hundreds of millions of dollars on bone marrow transplants and lung transplants and kidney transplants and heart transplants. These therapies are extraordinarily expensive, and they certainly don't save money for the health care system or society. But when we know a person who is sick, we'll stop at nothing to help. Unfortunately, we seem perfectly willing to withhold life-saving vaccines when we don't know who is going to be sick. We're willing to take that gamble—a gamble that many children will inevitably lose.

Religious and conservative groups also occasionally oppose vaccines. For example, the papillomavirus vaccine, capable of preventing the virus that causes cervical cancer, has become a political vaccine.

Every year in the United States, cervical cancer strikes about fifteen thousand women and kills five thousand. In 2006 the CDC recommended the papillomavirus vaccine for all adolescent girls before they become sexually active. Conservative groups feared that the vaccine would be mandated for all teenagers. They reasoned that papillomavirus is transmitted by a girl's having sex. If she doesn't have sex, she won't catch it. And if she and her partner don't have sex before marriage, then they won't be bringing papillomavirus into the marriage. The message should be abstinence before marriage, not prevention of a sexually transmitted disease by a vaccine that might only encourage sex. Tony Perkins was the president of the Family Research Council. He had no intention of vaccinating his thirteen-year-old daughter. "Our concern is that this vaccine will be marketed to a segment of the population that should be getting a message about abstinence," said Perkins. Leslie Unruh, leader of the National Abstinence Clearinghouse, agreed. "I personally object to vaccinating children against a disease that is one hundred percent preventable with proper sexual behavior." Art Caplan, director of the Center for Bioethics at the University of Pennsylvania, disagrees with Perkins and Unruh. "If you want to teach moral behavior to your children, teach it in the home or in the classroom or in the church or synagogue," he says. "But the notion of withholding a potentially lifesaving vaccine to promote moral behavior is unconscionable."

ANOTHER FORCE WORKING AGAINST VACCINES IS CAPITALISM. VACCINES aren't very profitable.

Unlike drugs that are often used every day, vaccines are used, at most, several times in a person's lifetime. So it isn't surprising that the market for drugs is substantially greater than that for vaccines. For example, the pneumococcal vaccine for children, the highest revenue-generating vaccine, had gross sales in 2006 of about $2 billion. But cholesterol-lowering agents, hair-loss products, sexual-potency drugs, or drugs for heart disease, obesity, or neurological problems can earn $7 billion or more per drug per year. Annual revenues for Lipitor, one of several cholesterol-lowering agents, were $13 billion—about twice the revenues for the entire worldwide vaccine industry.

Faced with these finances, companies have gradually abandoned vaccines. Twenty-six companies made vaccines in 1957; seventeen in 1980. Only five—GlaxoSmithKline, Sanofi-Pasteur, Merck, Pfizer, and Novartis—have a significant market presence today. (Many smaller companies also make vaccines, but their combined output accounts for less than 15 percent of the total market.) Within pharmaceutical companies, vaccines compete with drugs for resources—and lose. Because vaccines are made by only a few companies, problems with production often result in vaccine shortages. The influenza vaccine shortages from 2003 to 2005 were just one example. Since 1998, ten of sixteen vaccines routinely recommended for children have been in short supply—specifically, those to prevent measles, mumps, rubella, chickenpox, tetanus, diphtheria, pertussis, influenza, meningococcal infection, and pneumococcal infection. In 2002, Wyeth stopped making the DTaP and influenza vaccines. The decision to stop production had little impact on shareholders but a major impact on children. It precipitated shortages of and rationing for both vaccines. Some children never got the vaccines that they missed, even after the shortages were over.

Fortunately, the news about the vaccine business isn't all bad. Vaccines do have a couple of advantages over drugs. For example, no companies make generic vaccines for the United States market. So vaccines can continue to make a profit long after they're unprotected by patents. Also, unlike drugs, vaccines are recommended for routine use by agencies such as the CDC and the AAP, guaranteeing a market. Newer vaccines, such as those for rotavirus and meningococcus, are likely to make between $500 million and $1 billion per year and the papillomavirus vaccine could bring in revenues that exceed $2 billion. While revenues from vaccines are less than for blockbuster drugs, they are good enough to keep some pharmaceutical companies interested in continuing to make them and to invest in research to develop new ones.

IMAGINE WHAT WOULD HAPPEN IF SOME COMBINATION OF FORCES caused pharmaceutical companies to stop making vaccines or a critical number of people in the United States to stop using them.

The first disease to appear would be whooping cough (pertussis). Although most people don't realize it, pertussis occurs commonly in the United States. Every year it causes people to suffer prolonged bouts of coughing that last as long as several weeks. About one million teenagers and adults are infected. Adults don't die of pertussis. But babies do. And infants usually catch pertussis from infected adults. Pertussis bacteria cause inflammation in their small windpipes. Sometimes the inflammation is so severe that babies simply stop breathing. If the pertussis vaccine weren't available, babies would almost immediately start to die of pertussis.

Before the pertussis vaccine was first introduced in the 1940s, pertussis killed as many as eight thousand babies every year in the United States. Some people might argue that the 1940s were a long time ago and that we now have at our disposal antibiotics and intensive care facilities that we didn't have then. But antibiotics don't treat pertussis; they only stop transmission of the bacteria, making people with the disease less contagious, not less sick. When immunization rates in England dropped from 80 percent to 30 percent in the late 1970s, more than one hundred thousand children with pertussis were hospitalized, and about seventy died. And that was all within a couple of years of a fifty-point decrease in immunization rates.

Within a few years, measles, which has been virtually eliminated from the United States, would again find a home. Although measles infections are very rare in the United States, they're not rare in the world; every year about one million people are infected with the virus. Because international travel is common, people with measles often come into the United States. For example, in May 2006 an unimmunized computer programmer brought measles with him from India. He worked at the John Hancock Tower, Boston's tallest building and home to five thousand other workers. Anita Barry, Boston's public health commissioner, remembered what happened when the city first offered vaccine to company workers: "Only about thirty people showed up at that initial clinic," she said. "I wish that a lot more people had shown up. And so did the company, given subsequent events." Soon the virus spread to others in the building. Then it traveled a few blocks, infecting a member of the Christian Science Church, a group that doesn't believe in immunizations. "It's actually more contagious than smallpox," said Alfred DeMaria, Massachusetts's chief infectious diseases official. When fourteen more people were infected and the media alerted Boston residents of the growing outbreak in the Hancock Tower, thousands of people were vaccinated, and the outbreak subsided. But imagine what would have happened if no one in the John Hancock Tower had been immunized.

When enough people are unimmunized, rubella would again cause birth defects and fetal death. And mumps would be a common cause of deafness. In 2006 a mumps epidemic swept across several Midwestern states. The virus infected four thousand people, mostly young adults, causing seizures, meningitis, and deafness in about thirty people. The strain of mumps virus that was isolated in the United States was the same as the one recently found circulating in England, where seventy thousand people were infected. The outbreak in England was worse because many more people were unimmunized, scared by the false concern that the MMR vaccine caused autism.

Next up would be the bacterium *Haemophilus influenzae* type b (Hib). Before the vaccine, Hib caused meningitis and bloodstream infections in tens of thousands of children every year. And the Hib vaccine wasn't introduced in the United States until the mid-1980s, well after several antibiotics were available to treat the disease. Despite the availability of these antibiotics, Hib caused developmental delay, speech disorders, language delays, hearing deficits, and paralysis in many of those infected. Although antibiotics save lives, they aren't given until after the infection has already started, often too late to prevent permanent harm.

After about ten years in which babies were no longer immunized, polio would return, allowing younger Americans to experience first-hand what their parents and grandparents had experienced in the 1940s and 1950s. Again, parents would be scared to let their children go swimming in the summer or drink from water fountains or congregate in movie theaters or play with neighbors. This scenario isn't so far-fetched. In 1978 and again in 1992, outbreaks of polio occurred among members of a Dutch Reformed Church in the Netherlands who refused vaccines, causing paralysis in several children. Fortunately, the disease didn't spread to neighbors because 98 percent of people who came in contact with these children had been immunized. But if immunization rates had been only 50 percent or 0 percent, the outcome would have been very different.

When there were enough susceptible children, diphtheria would return. In the early 1990s, during the collapse of the Soviet Union, many children there didn't get the vaccines they needed. It wasn't long before diphtheria reappeared, causing disease in fifty-thousand people. The bacterium that causes diphtheria is still out there; if we lower our defenses, the disease will be back. And we will again be able to experience what it was like for our ancestors, who, in the 1920s, feared diphtheria as one of the greatest killers of teenagers.

Given other advances in modern medicine, some might argue

that we are better able to handle the onslaught of infectious diseases now than we were fifty years ago. But in some ways we're worse off. Intensive care units in hospitals today care for hundreds of cases of severe infections every year, not tens of thousands. Facilities to handle the sudden increase in infections that would inevitably follow a dramatic decline in immunization rates don't exist. Furthermore, intensive care units today take care of a much more vulnerable group of patients, such as severely premature babies and bone marrow or organ transplant recipients. Also, because of the widespread use of drugs like steroids, which suppress the immune system, many more people in the United States have weakened immune systems today than they did fifty years ago. As a consequence, they are far more susceptible to harm caused by highly contagious infections.

A return to a time before vaccines would also affect the workplace. Fifty years ago, most mothers stayed home to take care of their children. But today women work outside the home. If infectious diseases like measles, mumps, Hib, polio, and diphtheria returned, one parent would be forced to spend at least several more weeks every year staying home to take care of sick children. Day care centers, where spread of these diseases would be rapid, would be particularly vulnerable. The result would be a major disruption of the workforce necessary to provide goods and services, and billions of dollars in lost wages.

We don't have to go back in time to see what life was like before vaccines. If we want to get a peek into a future without vaccines, we could take a trip to sub-Saharan Africa or India or Pakistan and watch children suffer and die of diseases now prevented by vaccines, remembering that about half of these children are actually vaccinated. From measles alone, about two hundred thousand children die every year.

"Despite all of society's negative pressures, vaccination has proven itself beyond the shadow of doubt to be the most logical way to control infectious diseases in a community," says Adel Mahmoud, the parasitologist mentioned earlier who grew up in Egypt, a country devastated by infectious diseases. "The success story is undeniable. There is no measles, a little bit of mumps, no rubella, a little bit of hepatitis B in many communities. And the reason is vaccination. Vaccination is an unbelievably smart way of changing the environment for pathogens in human populations. It is as ecologically important as anything that we have discovered in our long history in the fight between us and the microbes. But it's not free. It comes with a price, an imperative. And that is that you have to keep using it."

# CHAPTER 11

## Unrecognized Genius

*One of these men is genius to the other;*
*And so of these. Which is the natural man,*
*And which the spirit? Who deciphers them?*

William Shakespeare, *Comedy of Errors*

"Here was a guy," said Walter Strauss, senior director of epidemiology research at Merck, "born on some windswept ranch in Montana, practically orphaned at birth, taken in by relatives, and who, but for his talent and drive, might have spent a lifetime working as a clerk at a retail store. Instead he rose to the pinnacle of scientific achievement in the United States, leaving his mark on half the world's children. It is one of the greatest of all Horatio Alger stories."

Hilleman didn't see his career as improbable, reasoning that his farming background was perfect training for a scientist to create life-saving vaccines. "We had a machine shop, an electrical shop, and a blacksmith shop," he said. "You learned agronomy. We tore apart irrigation pumps and put them back together. And I had an old 1928 Ford, nothing but a wreck. But I rebuilt [it] and made it a functioning car. When you're brought up on a farm, you have a lot of general knowledge." At Custer County High School Hilleman had the option of majoring in general farming, mechanics, science, business, or academics. He chose science. So did all of his brothers. And, like Hilleman, all were successful. Howard, a professor of anatomy and physiology at the University of Oregon, wrote a textbook on invertebrate anatomy. "He had a photographic memory," recalled Hilleman. Victor ran a landscaping unit for FDR's Civilian Conservation Corps and later built ships and airplanes. Harold designed and built propeller planes for Lockheed. Richard supervised the installation of electrical systems in planes, also for Lockheed. Norman was a radar specialist for the federal government. "He liked to memorize all of the radar circuitry," recalled Hilleman. The only brother who didn't choose a life in science or engineering was Walter, who graduated from Concordia College and was about to be ordained as a Lutheran minister when he died on the operating table of an undiagnosed case of appendicitis; he was nineteen years old.

Although all of his brothers were successful, Maurice's relentless, unending drive was unique. When asked to explain this difference, Hilleman would invariably talk about his father, saying "I wanted to get him to see me." Hilleman, raised by his aunt and uncle, hated his father for the perceived rejection. "I found [my father's] narrow-minded, domineering opinions and intrusions insufferable and unforgivable," recalled Hilleman. "The antagonism that I [felt] toward him was taken out on [my] occasional outbursts and his threats of physical harm. There was no resolution." But nothing ever erased the pain of getting up every day to watch his brothers and sister grow up in a house with their father—his father. Never recognized by his father, each of Hilleman's many accomplishments was just one spade of dirt in a bottomless hole.

As a child, Hilleman "cheated death." Neither his mother nor his twin sister survived his birth. If Hilleman hadn't survived, would we still have the same vaccines that we do today? Most likely all of his vaccines would have been developed eventually by others, with two exceptions: the blood-derived hepatitis B vaccine and the mumps vaccine. Only Maurice Hilleman had the foresight to treat human blood with chemicals, prove that he could kill any possible contaminating organism, and purify Australia antigen from the mix. No one else had the resources, intelligence, and guts to do this. With the appearance of HIV in the blood supply, the task was seen as too dangerous. But for the five years when Hilleman's blood-derived hepatitis B vaccine was the only one available to prevent hepatitis B virus infections, millions of doses were sold, and thousands of lives were saved. Also, no one has ever developed a better mumps vaccine than the one Hilleman made from the virus that infected his daughter. And it wasn't for lack of trying. In the 1960s a mumps vaccine made in Russia, called the Leningrad strain, and one made in Japan, called the Urabe strain, were licensed and sold. Both of these vaccines worked well to eliminate mumps infections and were on the market for decades. But both also came with a price: they occasionally caused meningitis. The Jeryl Lynn strain of mumps vaccine didn't cause this dangerous side effect.

Maurice Hilleman died on April 11, 2005. The day after his death his obituary appeared on the front page of the *New York Times*. Lawrence Altman wrote it. While writing his story, Altman asked prominent scientists and doctors why Hilleman wasn't better known to the public. Shock-jock Howard Stern read Altman's article and asked his listeners why they knew about Britney Spears's pregnancy but didn't know about Hilleman. That same day, public health officials, epidemiologists, clinicians, and members of the media gathered at the University of Pittsburgh's Alumni Hall to celebrate the fiftieth anniversary of Jonas Salk's polio vaccine. At a party following the ceremony, one of the participants told a group of pediatric infectious disease specialists that Maurice Hilleman had died. No single group of doctors was better positioned to appreciate the impact of Hilleman's work, but after hearing the news, all looked up with blank expressions, unmoved. Not one of them had ever heard his name.

Hilleman's relative anonymity can be explained in several ways.

Despite his self-confidence; profane, confrontational style; and domineering, occasionally frightening manner, Maurice Hilleman was a humble man.

When Anton Schwarz made his measles vaccine in the mid-1960s—one designed to compete with Hilleman's—he called it the Schwarz strain. When Jennifer Alexander took liver cells from a man dying from liver cancer, and later found that they produced Australia antigen, she called them Alexander cells. And when D. S. Dane looked through an electron microscope and saw hepatitis B virus particles circulating in human blood, he called them Dane particles. Hilleman was different; not one of his discoveries bears his name. He named his measles vaccine the Moraten strain (*Mor*e *At*tenuated *En*ders), in recognition of the work of John Enders. He named his rubella vaccine the HPV77-duck strain, in recognition of the *H*igh-*P*assage *V*irus originally developed by Harry Meyer and Paul Parkman. He named his two hepatitis B vaccines the plasma-derived and recombinant vaccines, noting the starting material and scientific process used to make them. After being the first to identify hepatitis A virus, he called the strain CR326. After finding that a previously unknown strain of monkey virus had contaminated early lots of polio vaccines, he called it simian virus-40. Hilleman allowed himself only one conceit: he named his mumps vaccine the Jeryl Lynn strain. But he didn't name it the Jeryl Lynn Hilleman or JLH strain. And few people today reading the package insert know that Jeryl Lynn was Maurice Hilleman's daughter.

"He was interested in the result and the product, not in taking credit for [it]," recalled Tony Fauci, director of the National Institute of Allergy and Infectious Diseases within NIH. "When he had a vaccine or a discovery, his attitude was more, 'Isn't this an interesting discovery,' rather than, 'I, Maurice Hilleman, did this.' It's almost like it never crossed his mind. He didn't really care about that. He just did [the work] and he let his accomplishments do the talking. So people know an incredible amount about what he did, but they don't know that it was he who did it. And when the obituaries and the eulogies came out, that's when people said, 'Oh, my God, this one guy did all of this?'"

"Despite his astounding accomplishments," recalled Walter Strauss, "Maurice carried himself with great humility. [We all] know many scientists with enormous egos that rest upon the smallest achievements. Maurice was different. Coming from a hardscrabble childhood in the Depression, he realized how much he had to be thankful for. Many of his childhood friends no doubt spent their adult lives working Montana cattle farms, wearing their bodies down for little reward. Maurice looked back on that alternative and considered himself tremendously lucky to be able to devote himself to something that was so much fun."

Because most people view industry researchers as being different from academic researchers, Hilleman's choice to work for a pharmaceutical company also contributed to his anonymity. Scientists, teachers, and researchers in academia believe that they are pursuing a higher calling, free from the bonds of commercialism. The public believes it too: people want scientists to be so dedicated and idealistic that they can live on air. In 1902 Wilhem Conrad Röntgen won the first Nobel Prize in physics for his discovery of X-rays. Röntgen believed that scientific knowledge was "to be freely shared for the good of humanity," not rewarded with something as base and common as money. Röntgen gave the Nobel Prize money—seventy thousand gold francs—to charity. Twenty years later he died, penniless.

In the late 1950s Samuel Katz was part of John Enders's research team at Boston Children's Hospital, involved in the quest to make the first measles vaccine. The Enders group never patented its vaccine. In 2005, Robert Kennedy Jr., in his article for *Rolling Stone* magazine, accused Sam Katz of having profited from a patent on the original measles vaccine. Katz was incensed: "I am cited personally as having a patent for a measles vaccine. That is just a total lie. I was part of the group of three who developed measles vaccine and brought it to licensure in 1963. However, our leader and mentor, John Enders, was a scientist who believed fully that the more people who are able to work on a problem, the more rapidly and likely it will be solved. Therefore, throughout our more than seven years of research we gave freely to any legitimate investigator who came to our laboratories. Dr. Enders was firmly opposed to patenting a biological product such as a vaccine, and we absolutely did not." Katz wanted Kennedy to know that the Boston researchers would never have debased themselves by financially profiting from their work; they were academicians, guardians of the public good, not industrialists out for profit.

Jonas Salk also made it clear to the press and the public that he would not profit from his polio vaccine. In April 1955, on the television program *See It Now*, Edward R. Murrow asked Salk, "Who owns the patent on this vaccine?" Salk thought for a moment and said, "Well, the people, I would say. There is no patent. Could you patent the sun?"

As reflected in popular culture, today's press and public share the disdain of Röntgen, Katz, and Salk for scientists interested in financial gain. A 1996 movie, *Twister*, subtly mirrors this belief. *Twister* is the story of two rival research groups trying to understand the physics of tornadoes. Each hopes that by placing small robotic measuring devices in the center of tornadoes it can better predict when and where the next one will appear. One research team is from academia, the other from industry. The academia group, led by Bill Paxton and Helen Hunt, consists of men and women of European, African, and Asian descent; all wear bright, colorful clothing that collectively looks like an entertaining patchwork quilt. The industry group, led by actor Cary Elwes, consists of white men in dark clothing; they look like a platoon from Darth Vader's home ship. The academia group is funny, childlike, and irreverent. The industry group is serious, formal, and humorless. The implications are clear. Academic research is fun, pursued by those with the curiosity of children who are naïve and pure at heart; they seek knowledge because knowledge alone is rewarding. Industry research, on the other hand, is serious, pursued by grim, faceless adults; knowledge is obtained solely for the money it can bring. We are much more comfortable touting the accomplishments of scientists in academia than of those in industry. "[Maurice] got enormous peer recognition," said Robert Gallo, codiscoverer of HIV. "But do we know anyone as a scientist in the corporate world who became well known? I don't think so."

"When Maurice came to work at Merck, it was wonderful," recalled Maurice's wife, Lorraine. "You had people to wash your pipettes and to wash your glassware. You didn't have to do any of that yourself. He couldn't believe that. And there was money to spend to do what you needed to do. Money wasn't an object. You could do your research." But Hilleman understood that most people saw pharmaceutical company scientists as inferior to those who worked in academia. He sarcastically referred to his working for "dirty industry." But he also knew that with the resources provided by industry he could have an impact on human health that would never be matched by the greatest academic centers. And that was a trade he was perfectly willing to make.

Hilleman's choice to get a doctorate in microbiology, not a medical degree, also contributed to his lack of recognition. When Hilleman finished weakening his daughter's mumps virus in the laboratory, he asked Robert Weibel and Joseph Stokes Jr. to test it. After Weibel and Stokes published their findings, and children in the United States began to receive the vaccine, Merck distributed a heart-warming picture to help physicians understand the vaccine's origins. In the center of the picture is two-year-old Kirsten Hilleman, getting a shot of the new vaccine. Tears are coursing down her face, and her mouth is open in a wide circle, screaming. Her sister, Jeryl Lynn, stands to her right. "I was telling her that everything was going to be all right," remembers Jeryl. To Kirsten's left—giving the vaccine—is Robert Weibel. Merck made thousands of copies of this photograph and distributed it to media outlets throughout the United States and the world. Although both of his daughters were in the photograph, Hilleman was nowhere to be found. Many people interpreted this photograph to mean that Robert Weibel had developed the mumps vaccine. Because Hilleman was a PhD working behind the scenes—not an MD on the front lines, giving vaccines and explaining them to the press and the public—few knew his name or his contributions.

Hilleman's lack of recognition didn't end with the mumps vaccine. After Wolf Szmuness completed the trial of Hilleman's blood-derived hepatitis B vaccine, he published his findings in one of medicine's most prestigious journals, the *New England Journal of Medicine*. Newspapers and magazines declared the importance of Szmuness's work. Radio stations interviewed him about his findings. Television stations showed video clips of Szmuness inoculating men who had volunteered for the trial—footage provided by Merck. When professional societies held meetings to discuss the hepatitis B vaccine, they called Wolf Szmuness. When advisory bodies wanted to determine exactly how the hepatitis B vaccine should be used in the United States, they called on Szmuness for answers. As far as the press, the public, and public health agencies were concerned, Wolf Szmuness had developed the hepatitis B vaccine.

But Szmuness hadn't developed the hepatitis B vaccine. Maurice Hilleman had. Szmuness's seminal publication in the *New England Journal of Medicine* contained the names of nine authors; Hilleman's wasn't among them. As a consequence, the press didn't seek Hilleman out, and doctors, nurses, and public health officials had no idea that he was the inventor. "I wanted to stand back and let Wolf determine whether the vaccine worked or not," recalled Hilleman. "I thought that if my name appeared on the paper, or if I was the one put in front of the television cameras or radio microphones, people would think that I was selling something. Because I was in industry and it was an evaluation of my work, ethically I felt that I had to stand back." Because of his own reticence and his company's discomfort about promoting him, few people recognized Maurice Hilleman for what he considered to be his greatest accomplishment.

Hilleman's work on the measles vaccine has also been largely ignored. Between 1989 and 1991 measles virus reemerged in the United States; about ten thousand people were hospitalized, and more than one hundred were killed by the virus. In response to the epidemic, the CDC recommended that all children receive a second dose of measles vaccine during childhood. The recommendation worked. In 2005, federal advisors, vaccine makers, the media, and the public gathered in Atlanta to hear the results of the new recommendation. The CDC reported that during the previous year only thirty-seven new cases of measles had occurred; no one had been hospitalized, and no one had been killed by the virus. Public health officials were excited about the possibility of finally eliminating measles from the United States. During the meeting, the chairman of the federal advisory committee recognized Sam Katz as the developer of the measles vaccine. Katz was a liaison member of the committee. "I'd just like to take a moment to recognize the contributions of Sam Katz," said the chairman. "He is the man who gave us the measles vaccine." Katz received an ovation. A humble, honest man, Katz held his hands up and tried to quiet the audience. "That's enough," he said.

Sam Katz and the Boston team made an important contribution by isolating measles virus and weakening it in their laboratory. They deserved an ovation. But their vaccine wasn't weak enough; it caused high fever and rash in as many as four of every ten children who received it. A better vaccine was made by Maurice Hilleman, who, unfortunately, never received a standing ovation for his achievement.

For all of his accomplishments, Maurice Hilleman never won the Nobel Prize. Because the prize cannot be awarded posthumously, he never will win it.

Alfred Nobel decided to create his prize after reading his own obituary. Born on October 21, 1833, in Stockholm, Sweden, Nobel was interested in explosives—specifically those that could be used in coal mines. At the time, miners used black powder, a form of gunpowder. But Nobel was interested in a much more powerful, volatile explosive: nitroglycerin. In 1862 he built a small factory to make it. One year later, Nobel invented a practical detonator that consisted of a wooden plug inserted into a metal tube; the plug contained black powder, and the tube contained nitroglycerin. Two years later Nobel improved his detonator by replacing the wooden plug with a small metal cap and replacing the black powder with mercury fulminate. This second invention, called a blasting cap, ushered in the modern age of explosives.

Nitroglycerin, however, remained difficult to handle. In 1864 Nobel's nitroglycerin factory blew up, killing Nobel's younger brother

Emil as well as several others. Undaunted, Nobel continued to work with the volatile chemical. By 1867 he had found that he was able to stabilize nitroglycerin by adding dirt containing large quantities of silica. He called his third invention dynamite, from the Greek *dynamis*, meaning "power." Dynamite provided armies with a new deadly weapon and made Alfred Nobel a very rich man.

In 1888 Alfred's brother Ludvig died in Cannes. French newspapers—confusing Ludvig with Alfred—printed Alfred's obituary under the headline "Le marchand de la mort est mort" (The merchant of death is dead). Nobel had always believed that his explosives, apart from their practical value, would be used as weapons of peace, not war. "My dynamite will sooner lead to peace," he said "than a thousand world conventions. As soon as men find that in one instant whole armies can be utterly destroyed, they surely will abide by golden peace." (Where have we heard this before?) After reading "his" obituary, Nobel realized that he was wrong; he would be remembered as a war maker, not a peacemaker. So on November 27, 1895, Nobel revised his will. "The whole of my remaining real estate shall be dealt with in the following way: the capital shall be annually distributed in the form of prizes to those who, during the preceding year, shall have conferred the greatest benefit to mankind [and it will be] divided into five equal parts: physics, chemical discovery or improvement, physiology or medicine; the field of literature; and the best work for fraternity between nations and promotion of peace." One year after revising his will, on December 10, 1896, Alfred Nobel died of a stroke in San Remo, Italy. Five years later, the king of Sweden awarded the first Nobel Prize. Today, no award is more coveted.

In accordance with Nobel's will, scientists working at the Karolinska Institute in Stockholm determine the winner of the Nobel Prize in physiology or medicine. But scientists are often more enamored with technological innovations than with public health achievements. As a consequence, the most important discoveries in the field of medicine have not been given to those who have saved the most lives.

Hilleman's accomplishments in the field of vaccines weren't the only ones snubbed by the Nobel Prize committee. Jonas Salk's polio vaccine caused a dramatic decrease in the incidence of polio in the United States and eliminated polio from several other countries. Within days of the announcement that his vaccine worked, Salk was honored at the White House. During a ceremony in the Rose Garden, in a voice trembling with emotion, President Dwight Eisenhower said, "I have no words to thank you. I am very, very happy." But Jonas Salk never won the Nobel Prize. Months after Salk's death, Renato Dulbecco, winner of the Nobel Prize in medicine in 1975 for his work on viruses that cause cancer, wrote Salk's obituary for the scientific journal *Nature*. "For his work on polio vaccine, Salk received every major recognition available in the world from the public and governments. But he received no recognition from the scientific world—he was not awarded the Nobel Prize, nor did he become a member of the U.S. National Academy of Sciences. The reason is that he did not make any innovative scientific discovery." Although Salk's vaccine was one of the greatest and most anticipated public health achievements of the twentieth century, the Nobel Prize would never be his.

A few years later, Albert Sabin developed his polio vaccine. By 1991 Sabin's polio vaccine had eliminated polio from the Western Hemisphere and much of the world. By 2030, if Sabin's vaccine continues to be given in Pakistan and Afghanistan, it will likely eliminate polio from the face of the earth. Sabin was a genius. Recognized by his colleagues, he won many awards and honors, including induction into the National Academy of Sciences. The Sabin Vaccine Institute in Washington, D.C., stands as a permanent shrine to honor the man. But Albert Sabin, like Jonas Salk, never won the Nobel Prize. The only Nobel Prize for the development of a polio vaccine was given to the Boston research team of John Enders, Fred Robbins, and Tom Weller. And that's because it was the Enders team that figured out how to grow polio virus in cell culture, allowing both Salk and Sabin to make their vaccines.

Maurice Hilleman made his measles, mumps, and rubella vaccines by weakening viruses in laboratory cells. His efforts led to the virtual elimination of those diseases in the United States. But Hilleman wasn't the first person to figure out how to grow viruses in cell culture, and he wasn't the first person to find that human viruses could be weakened in animal cells. That person was Max Theiler. So it was Theiler, in 1951, who won the Nobel Prize.

Hilleman considered his hepatitis B vaccine to be his greatest single achievement. His blood-derived vaccine, made by purifying Australia antigen from the blood of homosexual men in the late 1970s, was a technological tour de force. But Hilleman didn't discover Australia antigen; Baruch Blumberg did. And it was Blumberg who won the Nobel Prize for his finding. (Many scientists believe that the prize should have been awarded to Alfred Prince, the first to realize that Australia antigen was part of hepatitis B virus.)

Hilleman did, however, perform one series of studies worthy of the prize: his interferon research. Hilleman had been the first to purify interferon, determine its biological properties, and propose a mechanism for its action. Months before his death, knowing that this was Hilleman's last chance, several scientists lobbied members of the Nobel Prize committee, holding up Hilleman's interferon work as worthy of the prize. But one key member of the committee pointed out that the Nobel Prize in medicine would not be given to anyone who worked for a company.

Adding to the frustration of failed attempts to gain for Hilleman the recognition he deserved, the Nobel Prize committee has occasionally honored research that was far from deserving. For example, in 1926 the committee awarded the Nobel Prize in medicine to Johannes Fibiger for his discovery of a worm that caused stomach cancer in rats, figuring that this was an important breakthrough in what causes human cancer. But worms don't cause cancer in people. Julius Wagner-Jauregg won the prize in 1927 for discovering that malaria parasites could be used to treat syphilis; apart from being useless, the therapy was dangerous. Finally, in 1949 the committee awarded the prize to Egas Moniz of Portugal for discovering the value of lobotomy—the removal of a lobe of the brain—in treating certain psychoses. In the middle of the twentieth century, lo-

*Maurice Hilleman receives the National Medal of Science from President Ronald Reagan, 1988.*

botomies were popular. Doctors performed a lobotomy on John F. Kennedy's sister Rosemary; Ken Kesey had one performed on the fictional Randle P. McMurphy in his book *One Flew over the Cuckoo's Nest* (Jack Nicholson played McMurphy in the 1975 movie); and the *New England Journal of Medicine* hailed this therapy as the birth of "a new psychiatry." But the procedure proved worthless and cruel.

ALTHOUGH NOT RECOGNIZED BY THE PUBLIC, THE PRESS, OR THE Nobel Prize committee, Hilleman was honored by his colleagues. In 1983 he received the Albert Lasker Medical Research Award; in 1985 he was elected to the National Academy of Sciences; in 1988 he received the National Medal of Science from President Ronald Reagan; in 1989 he received the Robert Koch Gold Medal; in 1996 he received a Special Lifetime Achievement Award by the WHO;

and in 1997 he received the Albert B. Sabin Lifetime Achievement Award. "Every once in a while you come across a scientist whose list of accomplishments shine so brightly that you're almost blinded by them," recalled the NIH's Tony Fauci. "Most scientists would have been thrilled to have achieved just one of the scores and scores of Maurice's accomplishments."

Although he never received the Nobel Prize, because of Maurice Hilleman, hundreds of millions of children get to live their lives free from infections that at one time might have permanently harmed or killed them. In the final accounting, no prize is greater than that.

# Epilogue

*We remember the tale of the wonderful one-hoss shay*
*That was built in such a marvelous way*
*It lasted for a hundred years to the day*
*Then collapsed, all at once, in total decay*

Oliver Wendell Holmes
(Maurice Hilleman included this poem at the beginning of an autobiography that was forty pages long at the time of his death)

In 1984, when Maurice Hilleman was sixty-five years old, Merck, in accordance with company policy, asked him to retire. He refused. After months of negotiation, Merck, for the first time in its history, relented, allowing Hilleman to direct the newly created Merck Institute for Vaccinology. For the next twenty years Hilleman came to work every day, often staying late. There, he pored through recent scientific publications and wrote seminal review articles and opinion pieces about bioterrorism, the history of biowarfare, pandemic influenza, the vaccine enterprise, the history of vaccines, and the constant war waged between human beings and microbes. Not many scientists publish into their mid-eighties, but Hilleman's publications after his retirement continued to shape the way that scientists thought about vaccines and the diseases they prevented.

During his last twenty years, Maurice welcomed scientists from his company and the world, all seeking the wisdom of his experience.

On January 26, 2005, three months before his death, members of the scientific and medical community came to Philadelphia to honor Maurice Hilleman and to celebrate his life and work. They gathered at the American Philosophical Society.

Founded by Benjamin Franklin in 1743 and housed behind the white marble façade of the old Farmers and Mechanics Bank, the American Philosophical Society is the oldest surviving scholarly society in the United States. (In the eighteenth century "natural philosophy" meant the study of nature. If founded today, it would have been called the American Scientific Society.) Members of the society included founding fathers such as George Washington, Alexander Hamilton, John Adams, Thomas Jefferson, Thomas Paine, Benjamin Rush, James Madison, and John Marshall as well as scientists such as John Audubon, Robert Fulton, Thomas Edison, Louis Pas-

*Maurice Hilleman with daughters Kirsten (far left) and Jeryl and wife Lorraine, Vail, Colorado, December 1982. Maurice is the only one not wearing skis.*

teur, Albert Einstein, Linus Pauling, Margaret Mead, Marie Curie, and Charles Darwin; the society's library houses a first edition of Hilleman's beloved *The Origin of Species*.

Some of the best scientists of the twentieth century came to honor Hilleman that day. Robert Gallo (co-discoverer of HIV), Anthony Fauci (director of the National Institutes of Allergy and Infectious Diseases), Erling Norrby (a member of the Nobel Prize Committee), Hilary Koprowski (the co-developer of the modern rabies vaccine), Thomas Starzl (pioneer of liver transplantation), Roy Vagelos (developer of cholesterol-lowering agents), and Margaret Liu (an early pioneer of DNA vaccines) all stood up to talk about how Hilleman's accomplishments had instructed or dwarfed their own. At the end of the evening Hilleman thanked those in attendance. In a quiet, tired voice, he spoke of his friends, his family, and his good fortune. But Hilleman never forgot where he came from, recalling at the end of the symposium that although it was a wintry night on the streets of Philadelphia, "it was forty below in Montana."

Some in the audience cried, lamenting the end of a phenomenon.

DURING THE SYMPOSIUM, HILLEMAN, WHO LIKED TO THINK OF HIMSELF as tough and unmovable, said, "The most apt description of me was by Roy Vagelos, who said that on the outside I appeared to be a bastard but that if you looked deeper, inside, you still saw a bastard." But Hilleman's rough exterior hid a generous, softer side; he chose a career that benefited people because he loved people. When a friend at the CDC Foundation was diagnosed with breast cancer, he devoted himself to making sure that she got the care she needed. When neighborhood children asked to interview him for their science projects, he always complied; often spending hours carefully explaining his work. He was "the warmest family-oriented person that I've ever met," said Vagelos. "As tough as he was on people," recalled Fauci, "he did it in a way that was clear that he was trying to get a message to people and not to put anybody down. He just wanted straight talking. And when people veered from straight talking, he did not hesitate to get up and say, 'You're full of crap.' He was an ador-

*Maurice Hilleman shows grandson Dashiell a laboratory plate containing bacteria grown from Dashiell's diaper, 2004.*

able grump." His daughter Kirsten agreed. "People asked me what it was like to have Maurice as a father," she said. "It was wonderful. When we were little, our father traveled for weeks at a time and was a workaholic, but, strangely, I have almost no recollection of those long absences. What I remember best are those great stories he told about growing up in Montana; him singing me to sleep every night [with *The Sound of Music* song] 'Edelweiss.' And the many long hours we spent together in the basement conducting scientific experiments and building models. We built hearts, feet, hands, eyeballs, skeletons, transistor radios, and even a Model T Ford. My dad prided himself most on being a cynic, and yet he has accomplished so much in his life. His vaccines have saved so many lives that I believe that deep down under all of that crusty humor, he is an eternal optimist." "[He] taught me virtually everything that I knew," said Jeryl. "I learned to trust in logic and science while still leaving room for the mystical. My father gave me the greatest gift anyone can give another person. He believed in me."

At the symposium to honor her husband, Lorraine recalled their forty-two-year marriage. "Many of you know how we met," she said. "He was interviewing me [to coordinate] studies being done at Children's Hospital of Philadelphia for clinical trials of the measles and mumps vaccines. The interview was going very well, and he was pleasant enough. And suddenly he said, 'How old are you?' I told him I was twenty-nine. And he said, 'Oh, I thought you were thirty-five.' In spite of that we were married a short time later. And now for almost forty-two years he truly is a marvel of a man. And I am his lady."

DURING THE FEW MONTHS BEFORE HIS DEATH, HILLEMAN TALKED A great deal about the two men who had influenced his choices: his father, Gustave, and the man who raised him: his father's brother, Robert. "Gustave was not a bad man," recalled Hilleman. "But he was changed at the time of [my mother's] death and the shock it brought and the new responsibilities it entailed. He became newly charged with religious zeal and his need to get his family to heaven, best accomplished by having all his boys become ministers and his daughter marrying one. He sang loudly and prayed from his pew in the eighth row and continuously proclaimed that 'the Lord would provide,' no matter what adversity would befall him. To provide medical care for his family, he decided that he should enroll in a correspondence school to become a chiropractor. In chiropractic belief at that time it was claimed that all disease is caused by an impingement on the spine. Automatically, this threw out the infectious germ theory for disease and, above all, prohibited prevention by vaccination. I was [not allowed] to have any vaccines." (Perhaps this is one of medicine's greatest ironies.)

Hilleman's adoptive father, Robert, was different. "Bob was the antipathy and opposite of Gustave: freethinking, liberal, considerate, and, above all, highly intelligent. Bob's religion was practiced in good deeds. He would order up and deliver goods from out of state—margarine and other products not allowed by the dairy lobby. Everything was at cost, no profit. He sold affordable Lutheran life insurance to cover minimal family needs. And he provided informed legal assistance and wrote wills for those who couldn't pay for lawyers."

Faced with his imminent death, Hilleman thought a lot about these two men. At the end, he didn't turn to the religion of his father. Rather, as he had throughout his life, he sought solace in the reason and freethinking of his uncle. The ultimate experimentalist, Hilleman wanted to see whether one particular theory about cancer vaccines worked. So he tried it on himself.

In the 1940s researchers showed that if malignant tumor cells were taken from one mouse and injected into another, they caused cancer in the second mouse. But if the malignant tumor cells were weakened and injected into mice and then the same mice were injected with virulent cancer cells, they wouldn't get cancer. Mice could be vaccinated against cancer. For a long time it wasn't clear why this happened. Then a graduate student named Pramod Srivastava, now a professor of immunology at the University of Connecticut, figured it out. He took tumor cells, broke them open, separated different parts of the cells, and found that one particular group of proteins, called heat-shock proteins, protected mice against cancer. Heat-shock proteins are usually found in cells under stress caused by heat, cold, lack of oxygen, or cancer. Srivastava found that heat-shock proteins—with small fragments of cancer proteins in tow—were trying to warn the immune system that a cancer was growing. Srivastava reasoned that if he could take the cancer, purify the heat-shock proteins containing cancer proteins, and inject large quantities of this combination back into mice, then he could eliminate the cancer. Researchers working with many different cancers found that this worked in mice and rats. But they didn't know if it worked in people. When Hilleman wanted to see if heat-shock proteins could eliminate his own cancer, studies in people were under way, but none had been completed.

Hilleman sent his cancer cells, taken from the space between his lungs and his chest wall, to a company that specialized in purifying heat-shock proteins. He was hoping that he would live long enough to inject himself with 25 micrograms of these proteins once a week for four weeks. But Hilleman's cancer overwhelmed him before he could complete his final experiment.

. . .

On April 14, 2005, Maurice Hilleman was finally laid to rest near his home in Chestnut Hill, Pennsylvania. "In Montana, when there was something to celebrate," recalled Hilleman, "everybody would get together, sit on a log, get a fresh bucket of water, and pass around a cup." Hilleman had watched this scene many times. But Maurice Hilleman never rested, never celebrated, never stood back to enjoy what he had done. He was always driven to make the next discovery. "We got another license [for a vaccine]," he recalled. "So what. We're not going to sit around and celebrate. We've got new vaccines to make. You've got a bunch of mountains. When you climb one, then you've got another one to climb and another one. When you get to the top of one mountain, interest declines. There's more work to be done."

One hundred years from now we'll open the National Millennium Time Capsule commissioned by Bill and Hillary Clinton. We'll find the clear plastic block containing Maurice Hilleman's vaccines. We'll also find an artifact from Peggy Prenshaw, winner of the National Humanities Medal. Prenshaw submitted the words of author William Faulkner when he accepted the Nobel Prize for literature in 1950: "I believe that man will not merely endure: he will prevail. He is immortal, not because he alone among creatures has an inexhaustible voice, but because he has a soul, a spirit capable of compassion and sacrifice and endurance." Maurice Hilleman represented that spirit: a man whose work was unprecedented and whose gifts will remain forever, unmatched.

# Notes

This book is based on a series of interviews conducted with Dr. Maurice R. Hilleman on November 12, 19, and 30, 2004; December 3, 10, 15, and 17, 2004; January 6, and 7, 2005; February 17, 2005; and March 11, 2005.

### Prologue

The history of vaccine development is contained in Plotkin and Orenstein, *Vaccines,* and Plotkin and Fantini, *Vaccinia.*

### The Time Capsule

xxiii National Millennium Time Capsule: T. Stephens, "UCSC Researchers Produce Human Genome CD for the National Millennium Time Capsule," *UC Santa Cruz Currents Online,* January 22, 2001, www.ucsc.edu/currents; Press Release, National Archives and Records Administration, "National Millennium Time Capsule Exhibition to Open at the National Archives," Press release, December 4, 2000, www.archives.gov/mediadesk/pressreleases/nr01–20.html; CNN.com: "Nation's Time Capsule: Dog Tags, a Cell Phone and Dreams," December 6, 2000, http://archives.cnn.com/2000/US/12/06/timecapsule.ap; and "Clintons Busy with Duties before Dawn of 2000," December 31, 1999, http://archives.cnn.com/1999/ALLPOLITICS/stories.12/31.kickoff; White House Millennium Council: "National Millennium Time Capsule," http://clinton3.nara.gov/Initiatives/Millennium/capsule.html, and http://clinton4.nara.gov/Initiatives/Millennium/capsule/ theme_medalist.html; "Take This Capsule, Call in 100 Years," *New York Times*, January 2, 2000.

xxv Vaccine impact: J. P. Bunker, H. S. Frazier, and F. Mosteller, "Improving Health: Measuring Effects of Medical Care," *Milbank Quarterly* 72 (1994): 225–58.

**"My God: This Is the Pandemic. It's Here!"**

The best single description of the origin of bird flu in Southeast Asia is contained in Kolata, *Flu*.

1 Bird flu: E. Check, "WHO Calls for Vaccine Boost to Prepare for Flu Pandemic," *Nature* 432 (2004): 261; T. T. Hien, M. de Jong, and J. Farrar, "Avian Influenza: A Challenge to Global Health Care Structures," *New England Journal of Medicine* 351 (2004): 2363–65; K. Stöhr, and M. Esveld, "Will Vaccines Be Available for the Next Influenza Pandemic?" *Science* 306 (2004): 2195–96; R. S. Nolan, "Future Pandemic Most Likely Will Be Caused by Bird Flu," *AAP News*, January 2005; M. T. Osterholm, "Preparing for the Next Pandemic," *New England Journal of Medicine* 352 (2005): 1839–42; K. Ungchusak, P. Auewarakul, S. Dowell, et al., "Probable Person-to-Person Transmission of Avian Influenza A (H5N1)," *New England Journal of Medicine* 352 (2005): 333–40; A. S. Monto, "The Threat of Avian Influenza Pandemic," *New England Journal of Medicine* 352 (2005): 323–25; H. Chen, G. J. D. Smith, and S. Y. Zhang, "H5N1 Virus Outbreak in Migratory Waterfowl," *Nature* 436 (2005):191; J. Liu, H. Xiao, F. Lei, et al., "Highly Pathogenic H5N1 Influenza Virus Infection in Migratory Birds," *Science* 309 (2005): 1206; T. R. Maines, X. H. Lu, S. M. Erb, et al., "Avian Influenza (H5N1) Viruses Isolated from Humans in Asia in 2004 Exhibit Increased Virulence in Mammals," *Journal of Virology* 79 (2005): 11788–800; "The Next Killer Flu: Can We Stop It? *National Geographic*, October 2005; "Avian Flu: Ready for a Pandemic? *Nature*, May 26, 2005; D. Normile: "Outbreak in Northern Vietnam Baffles Experts," *Science* 308 (2005): 477, and "Genetic Analyses Suggest Bird Flu Virus Is Evolving," *Science* 308 (2005): 1234–35; D. Cyranoski, "Flu in Wild Birds Sparks Fears of Mutating Virus," *Nature* 435 (2005): 542–43; D. Butler: "Bird Flu: Crossing Borders," *Nature* 436 (2005): 310–11; "Flu Officials Pull Back from Raising Global Alert Level," *Nature* 436 (2005): 6–7; "Alarms Ring over Bird Flu Mutations," *Nature* 439 (2006): 248–49; "Doubts over Source of Bird Flu Spread," *Nature* 439 (2006): 772; "Yes, But Will It Jump?" *Nature* 439 (2006): 124–25; M. Enserink, "New Study Casts Doubt on Plans for Pandemic Containment," *Science* 311 (2006): 1084.

2 1918 influenza pandemic: Barry, *Great Influenza*; Kolata, *Flu*; R. B. Belshe, "The Origins of Pandemic Influenza: Lessons from the 1918 Virus," *New England Journal of Medicine* 353 (2005): 2209–11; J. S. Oxford, "Influenza A Pandemics of the 20th Century with Special Reference to 1918: Virology, Pathology and Epidemiology," *Reviews in Medical Virology* 10 (2000): 110–33.

3 AIDS pandemic: "The Global HIV/AIDS Pandemic," *Morbidity and Mortality Weekly Report*, April 11, 2006.

5 Jeryl Hilleman quote: "The Vaccine Hunter," BBC Radio 4, producer Pauline Moffatt, June 21, 2006.

9 Ricketts: H. T. Ricketts, "The Transmission of Rocky Mountain Spotted Fever by the Bite of the Wood-Tick (*Dermacentor occidentalis*)," *Journal of the American Medical Association* 47 (1906): 358.

**10** Lorraine Hilleman quote: "The Vaccine Hunter," BBC Radio 4, producer Pauline Moffatt, June 21, 2006.

**11** Infection and wars: McNeill, *Plagues*.

**13** *New York Times* article: "Hong Kong Battling Influenza Epidemic," *New York Times*, April 17, 1957.

**15** 1957 pandemic: S. F. Dowell, B. A. Kupronis, E. R. Zell, and D. K. Shay, "Mortality from Pneumonia in Children in the United States, 1939 through 1996," *New England Journal of Medicine* 342 (2000): 1399–1407; J. R. Schäfer, Y. Kawaoka, W. J. Bean, et al., "Origin of the Pandemic 1957 H2 Influenza A Virus and the Persistence of Its Possible Progenitors in the Avian Reservoir, *Virology* 194 (1993): 781–88; N. J. Cox, and K. Subbarao, "Global Epidemiology of Influenza: Past and Present," *Annual Reviews of Medicine* 51 (2000): 407–21; Y. Karaoka, S. Krauss, and R. G. Webster, "Avian-to-Human Transmission of the PB1 Gene of Influenza A Viruses in the 1957 and 1968 Pandemics," *Journal of Virology* 63 (1989): 4603–8; L. Simonsen, M. J. Clarke, L. B. Schonberger, et al., "Pandemic versus Epidemic Influenza Mortality: A Pattern of Changing Age Distribution," *Journal of Infectious Diseases* 178 (1998): 53–60; D. F. Hoft, and R. B. Belshe, "The Genetic Archaeology of Influenza," *New England Journal of Medicine* 351 (2004): 2550–51.

**18** Hilleman influenza studies: M. R. Hilleman, R. P. Mason, and N. G. Rogers, "Laboratory Studies on the 1950 Outbreak of Influenza," *Public Health Reports* 65 (1950): 771–77; M. R. Hilleman, R. P. Mason, and E. L. Buesher, "Antigenic Pattern of Strains of Influenza A and B," *Proceedings of the Society for Experimental Biology and Medicine* 75 (1950): 829–35; M. R. Hilleman, E. L. Buescher, and J. E. Smadel, "Preparation of Dried Antigen and Antiserum for the Agglutination-Inhibition Test for Influenza Virus," *Public Health Reports* 66 (1951): 1195–1203; M .R. Hilleman, "System for Measuring and Designating Antigenic Components of Influenza Viruses with Analyses of Recently Isolated Strains," *Proceedings of the Society for Experimental Medicine and Biology* 78 (1951): 208–15; M. R. Hilleman, "A Pattern of Antigen Variation," *Federation Proceedings* 11 (1952): 798–803; M. R. Hilleman, and F. L. Horsfall, "Comparison of the Antigenic Patterns of Influenza A Virus Strains Determined by in ovo Neutralization and Hemagglutination-Inhibition," *Journal of Immunology* 69 (1952): 343–56; M. R. Hilleman, and J. H. Werner, "Influence of Non-Specific Inhibitor of the Diagnostic Hemagglutination-Inhibition Test for Influenza," *Journal of Immunology* 71 (1953): 110–17; M. R. Hilleman, J. H. Werner, and R. L. Gauld, "Influenza Antibodies in the Population in the USA," *Bulletin of the World Health Organization* 8 (1953): 613–31; M. R. Hilleman, "Antigenic Variation of Influenza Viruses," *Annual Review of Microbiology* 8 (1954): 311–32; H. M. Meyer, M. R. Hilleman, M. L. Miesse, et al., "New Antigenic Variant in Far East Influenza Epidemic, 1957," *Proceedings of the Society for Experimental Medicine and Biology* 95 (1957): 609–16; M. R. Hilleman, "Asian Influenza: Initial Identification of Asiatic Virus and Antibody

Response in Volunteers to Vaccination," *Proceedings of a Special Conference on Influenza*, U. S. Department of Health, Education, and Welfare, August 27–28, 1957; M. R. Hilleman, F. J. Flatley, S. A. Anderson, et al., "Antibody Response in Volunteers to Asian Influenza Vaccine," *Journal of the American Medical Association* 166 (1958): 1134–40; M. R. Hilleman, F. J. Flatley, S. A. Anderson, et al., "Distribution and Significance of Asian and Other Influenza Antibodies in the Human Population," *New England Journal of Medicine* 258 (1958): 969–74; C. C. Mascoli, M. B. Leagus, and M. R. Hilleman, "Influenza B in the Spring of 1965," *Proceedings of the Society for Experimental Biology and Medicine* 123 (1966): 952–60; M. R. Hilleman, "The Roles of Early Alert and of Adjuvant in the Control of Hong Kong Influenza by Vaccines," *Bulletin of the World Health Organization* 41 (1969): 623–28.

18 Hilleman prediction about bird flu: M. R. Hilleman, "Realities and Enigmas of Human Viral Influenza: Pathogenesis, Epidemiology and Control," *Vaccine* 20 (2002): 3068–87.

**Jeryl Lynn**

Robert Weibel, Art Caplan, and Jeryl Lynn Hilleman were interviewed on January 6, 2005; March 10, 2005; and March 11, 2005, respectively.

21 Mumps disease: Plotkin and Orenstein, *Vaccines*.

22 Hilleman mumps vaccine studies: E. B. Buynak, and M. R. Hilleman, "Live Attenuated Mumps-Virus Vaccine, I: Vaccine Development," *Proceedings of the Society for Experimental Biology and Medicine* 123 (1966): 768–75; J. Stokes Jr., R. E. Weibel, E. B. Buynak, and M. R. Hilleman, "Live Attenuated Mumps-Virus Vaccine. II: Early Clinical Studies," *New England Journal of Medicine* 39 (1967): 363–71; R. W. Weibel, J. Stokes Jr., E. B. Buynak, J. E. Whitman, and M. R. Hilleman, "Live, Attenuated Mumps-Virus Vaccine; 3: Clinical and Serological Aspects in a Field Evaluation," *New England Journal of Medicine* 276 (1967): 245–51; M. R. Hilleman, R. W. Weibel, E. B. Buynak, J. Stokes Jr., J. E. Whitman, "Live, Attenuated Mumps-Virus Vaccine; 4: Protective Efficacy as Measured in a Field Evaluation," *New England Journal of Medicine* 276 (1967): 252–58; R. W. Weibel, E. B. Buynak, J. Stokes Jr., J. E. Whitman, and M. R. Hilleman, "Evaluation of Live Attenuated Mumps Virus Vaccine, Strain Jeryl Lynn," Pan American Health Organization, *Scientific Publication No. 147*, May 1967, 430–37; R. W. Weibel, J. Stokes Jr., E. B. Buynak, M. B. Leagus, and M. R. Hilleman, "Jeryl Lynn Strain Live Attenuated Mumps Virus Vaccine: Duration of Immunity Following Administration," *Journal of the American Medical Association* 203 (1968): 14–18; R. W. Weibel, E. B. Buynak, J. E. Whitman, M. B. Leagus, J. Stokes Jr., and M. R. Hilleman, "Jeryl Lynn Strain Live Attenuated Mumps Virus Vaccine: Duration of Immunity for Three Years Following Vaccination," *Journal of the American Medical Association* 207 (1969): 1667–70; R. W. Weibel, E. B. Buynak, J Stokes Jr., and M. R. Hilleman, "Persistence of Immunity Four Years Following Jeryl Lynn Strain of Live Mumps Virus Vaccine," *Pediatrics* 45 (1970): 821–26.

24 Polio vaccine studies: Carter, *Breakthrough*; J. Smith, *Patenting the Sun*; Oshinsky, *Polio*; interview with Donna Salk, February 10, 1999; J. A. Kolmer, G. F. Klugh, and A. M. Rule, "A Successful Method for Vaccination against Acute Anterior Poliomyelitis," *Journal of the American Medical Association* 104 (1935): 456–60; J. A. Kolmer, "Susceptibility and Immunity in Relation to Vaccination with Acute Anterior Poliomyelitis," *Journal of the American Medical Association* 105 (1935): 1956–62.

26 Willowbrook studies: Rothman and Rothman, *Willowbrook*; G. Rivera, *Willowbrook: A Report on How It Is and Why It Doesn't Have to Be That Way* (New York: Vintage Books, 1972); S. Krugman, and R. Ward, "Clinical and Experimental Studies of Infectious Hepatitis," *Pediatrics* 22 (1958): 1016–22; S. Krugman, "The Willowbrook Hepatitis Studies Revisited: Ethical Aspects," *Reviews of Infectious Diseases* 8 (1986): 157–62.

27 Fernald studies: N. Fost, "America's Gulag Archipelago," *New England Journal of Medicine* 351 (2004): 2369–70.

30 Impact of mumps vaccine: Plotkin and Orenstein, *Vaccines*.

30 Mumps versus Jeryl Lynn: A. Dove, "Maurice Hilleman," *Nature Medicine Supplement* 11 (2005): 52.

**Eight Doors**

An excellent summary of the history of vaccine development and of the historic events that influenced Hilleman's work can be found in the following: Plotkin and Fantini, *Vaccinia*; M. R. Hilleman: "Vaccines and the Vaccine Enterprise: Historic and Contemporary View of a Scientific Initiative of Complex Dimensions," *The Jordan Report* (2002); "Personal Reflections on Twentieth Century Vaccinology," *Southeast Asian Journal of Tropical Medicine and Public Health* 34 (2003): 244–48; and "Overview of Vaccinology in Historic and Future Perspective: The Whence and Whither of a Dynamic Science with Complex Dimensions," in *DNA Vaccines*, ed. H. C. J. Ertl (Plenum Publishers, London, 2003).

31 Jenner and the smallpox vaccine: Tucker, *Scourge*; Radetsky, *Invaders*; Williams, *Virus Hunters*; B. Moss, "Vaccinia Virus: A Tool for Research and Vaccine Development," *Science* 252 (1991): 1662–67; M. Radetsky, "Smallpox: A History of Its Rise and Fall, *Pediatric Infectious Disease Journal* 18 (1999): 85–93.

33 Smallpox vaccine problems: Tucker, *Scourge*.

34 Pasteur: Radetsky, *Invaders*; Williams, *Virus Hunters*; Debré, *Pasteur*; Geison, *Private Science*.

36 Pasteur vaccine problems: Plotkin and Orenstein, *Vaccines*.

36 Beijerinck: van Iterson, De Jong, and Kluyver, *Beijerinck*; Radetsky, *Invaders*; Williams, *Virus Hunters*; "Beijerinck and His 'Filterable Principle,'" *Hospital Practice* 26 (1991): 69–86.

38 Carrel: Williams, *Virus Hunters*; http://crishunt.8bit.co.uk/alexiscarrel.html; http://members.aol.com/amaccvpe/history.carrel.htm; http://www.whonamedit.com/doctor.cfm/445.html.

38 Goodpasture: Collins, *Goodpasture;* A. M. Woodruff, and E. W. Goodpasture, "The Susceptibility of the Chorio-Allantoic Membrance of Chick Embryos to Infection with Fowl-Pox Virus," American Journal of Pathology 7 (1931): 209–22; M. Burnet, "The Influence of a Great Pathologist: A Tribute to Ernest Goodpasture," *Perspectives in Biology and Medicine* 16 (1973): 333–47.

39 Theiler: Williams, *Virus Hunters*; Plotkin and Orenstein, *Vaccines*; H. H. Smith and M. Theiler, "The Adaptation of Unmodified Strains of Yellow Fever Virus to Cultivation in vitro," *Journal of Experimental Medicine* 65 (1937): 801–8; M. Theiler and H. H. Smith, "The Effect of Prolonged Cultivation in vitro upon the Pathogenicity of Yellow Fever Virus," *Journal of Experimental Medicine* 65 (1937): 767–86; T. N. Raju, "The Nobel Chronicles," *Lancet* 353 (1999): 1450; M. A. Shampo, and R. A. Kyle, "Max Theiler: Nobel Laureate for Yellow Fever Vaccine," *Mayo Clinic Proceedings* 78 (2003): 728.

40 Yellow fever vaccine and hepatitis: J. P. Fox, C. Manso, H. A. Penna, and M. Para, "Observations on the Occurrence of Icterus in Brazil Following Vaccination against Yellow Fever," *American Journal of Hygiene* 36 (1942): 68–116; M. V. Hargett and H. W. Burruss, "Aqueous-Based Yellow Fever Vaccine," *Public Health Reports* 58 (1943): 505–12; W. A. Sawyer, K. F. Meyer, M. D. Eaton, et al., "Jaundice in Army Personnel in the Western Region of the United States and Its Relation to Vaccination against Yellow Fever," *American Journal of Hygiene* 40 (1944): 35–107; L. B. Seeff, G. W. Beebe, J. H. Hoofnagle, et al., "A Serologic Follow-Up of the 1942 Epidemic of Post-Vaccination Hepatitis in the United States Army," *New England Journal of Medicine* 316 (1987): 965–70.

40 Enders, Weller and Robbins: Williams, *Virus Hunters*; Radetsky, *Invaders*; Weller, *Growing Pathogens*; J. F. Enders, T. H. Weller, and F. C. Robbins, "Cultivation of the Lansing Strain of Poliomyelitis Virus in Cultures of Various Human Tissues," *Science* 109 (1949): 85–87; T. H. Weller, F. C. Robbins, and J. F. Enders, "Cultivation of Poliomyelitis Virus in Cultures of Human Foreskin and Embryonic Tissues," *Proceedings of the Society of Experimental Biology and Medicine* 72 (1949): 153–55; F. C. Robbins, J. F. Enders, and T. H. Weller, "Cytopathogenic Effect of Poliomyelitis Virus in vitro on Human Embryonic Tissues," *Proceedings of the Society of Experimental Biology and Medicine* 75 (1950): 370–74.

42 Salk: J. Smith, *Patenting the Sun*; Carter, *Breakthrough*; Tony Gould, *A Summer Plague: Polio and its Survivors* (New Haven and London: Yale University Press, 1995).

42 The Cutter Incident: Offit, *Cutter*.

**The Destroying Angel**

Samuel Katz and Thomas Weller were interviewed on February 22, 2005, and November 10, 2005, respectively.

44 Measles: Plotkin and Orenstein, *Vaccines*; M. R. Hilleman, "Current Overview of the Pathogenesis and Prophylaxis of Measles with Focus on Practical Implications," *Vaccine* 20 (2002): 651–65.

45 Peebles: Williams, *Virus Hunters*.

46 Enders's laboratory measles studies: J. F. Enders and T. C. Peebles, "Propagation in Tissue Cultures of Cytopathogenic Agents from Patients with Measles," *Proceedings of the Society for Experimental Biology and Medicine* 86 (1954): 277–86; J. F. Enders, T. C. Peebles, K. McCarthy, M. Milovanovic, et al., "Measles Virus: A Summary of Experiments Concerned with Isolation, Properties, and Behavior," *American Journal of Public Health* 47 (1957): 275–82; S. L. Katz, M. Milovanovic, and J. F. Enders, "Propagation of Measles Virus in Cultures of Chick Embryo Cells," *Proceedings of the Society of Experimental Biology and Medicine* 37 (1958): 23–29; J. F. Enders, S. L. Katz, M. V. Milovanovic, and A. Holloway, "Studies on Attenuated Measles-Virus Vaccine, I: Development and Preparation of the Vaccine: Techniques for Assay of Effects of Vaccination," *New England Journal of Medicine* 263 (1960): 153–69; S. L. Katz, J. F. Enders, and A. Holloway, "Studies on an Attenuated Measles-Virus Vaccines, II: Clinical, Virologic and Immunologic Effects of Vaccine in Institutionalized Children," *New England Journal of Medicine* 263 (1960): 157–61; S. Krugman, J. P. Giles, and A. M. Jacobs, "Studies on Attenuated Measles-Virus Vaccine; VI: Clinical, Antigenic and Prophylactic Effects of Vaccine in Institutionalized Children," *New England Journal of Medicine* 263 (1960): 174–77; S. L. Katz, H. Kempe, F. L. Balck, et al., "Studies on Attenuated Measles-Virus Vaccine; VIII: General Summary and Evaluation of Results of Vaccination," *New England Journal of Medicine* 263 (1960): 180–84.

48 Hilleman measles studies: J. Stokes Jr., C. M. Reilly, M. R. Hilleman, and E. B. Buynak, "Use of Living Attenuated Measles-Virus Vaccine in Early Infancy," *New England Journal of Medicine* 263 (1960): 230–33; C. M. Reilly, J. Stokes Jr., E. B. Buynak, G. Goldner, and M. R. Hilleman, "Living Attenuated Measles-Virus Vaccine in Early Infancy: Studies of the Role of Passive Antibody in Immunization," *New England Journal of Medicine* 265 (1961): 165–69; J. Stokes Jr., M. R. Hilleman, R. E. Weibel, et al., "Efficacy of Live, Attenuated Measles-Virus Vaccine Given with Human Immune Globulin: A Preliminary Report," *New England Journal of Medicine* 265 (1961): 507–13; J. Stokes Jr., C. M. Reilly, E. B. Buynak, and M. R. Hilleman, "Immunologic Studies of Measles," *American Journal of Hygiene* 74 (1961): 293–303; J. Stokes Jr., R. E. Weibel, R. Halenda, C. M. Reilly, and M. R. Hilleman, "Studies of Live Attenuated Measles-Virus Vaccine in Man, I: Clinical Aspects," *American Journal of Public Health* 52 (1962): 29–43; M. R. Hilleman, J. Stokes Jr., E.B. Buynak, et al., "Studies of Live Attenuated Measles-Virus Vaccine in Man, II: Appraisal of Efficacy," *American Journal of Public Health* 52 (1962): 44–56; J. Stokes Jr., R. Weibel, R. Halenda, C. M. Reilly, and M. R. Hilleman, "Enders Live Measles-Virus Vaccine with Human Immune Globulin, I: Clinical Reactions," *American Journal of Diseases*

of *Children* 103 (1962): 366–72; M. R. Hilleman, J. Stokes Jr., E. B. Buynak, et al., "Enders' Live Measles-Virus Vaccine with Human Immune Globulin," *American Journal of Diseases of Children* 103 (1962): 373–79; M. R. Hilleman and H. Goldner, "Perspectives for Testing Safety of Live Measles Vaccine," *American Journal of Diseases of Children* 103 (1962): 484–95; M. R. Hilleman, J. Stokes Jr., E. B. Buynak, et al., "Immunogenic Response to Killed Measles Vaccine," *American Journal of Diseases of Children* 103 (1962): 445–51; J. Stokes Jr., M. R. Hilleman, R. E. Weibel, et al., "Persistent Immunity Following Enders Live, Attenuated Measles-Virus Vaccine Given with Human Immune Globulin," *New England Journal of Medicine* 267 (1962): 222–24; R. E. Weibel, J. Stokes Jr., R. Halenda, E. B. Buynak, and M. R. Hilleman, "Durable Immunity Two Years after Administration of Enders's Live Measles-Virus Vaccine with Immune Globulin," *New England Journal of Medicine* 270 (1964): 172–75; M. R. Hilleman, E. B. Buynak, R. E. Weibel, et al., "Development and Evaluation of the Moraten Measles Vaccine," *Journal of the American Medical Association* 206 (1968): 587–90; E. B. Buynak, R. E. Weibel, A. A. McLean, and M. R. Hilleman, "Long-Term Persistence of Antibody Following Enders' Original and More Attenuated Live Measles Virus Vaccine," *Proceedings of the Society of Experimental Biology and Medicine* 153 (1976): 441–43; M. R. Hilleman, "Current Overview of the Pathogenesis and Prophylaxis of Measles with Focus on Practical Implications," *Vaccine* 20 (2002): 651–65.

**48** Stokes and gamma globulin: A. M. Bongiovanni, "Joseph Stokes Jr.," *Pediatrics* 50 (1972): 163–64.

**49** Clinton Farms: J. Peet, "Edna Mahan Clung to Her Ideal," *Newark Star Ledger*, January 5, 2006.

**50** Peyton Rous and cancer causing viruses: Williams, *Virus Hunters*; Radetsky, *Invaders*; Francis Peyton Rous, http://www.geocities.com/galenvagebn/RousFP.html; Britannica Nobel Prizes, http://www.britannica.com/nobel/micro/511_71.html; P. Rous, "Transmission of a Malignant New Growth by Means of a Cell-Free Filtrate," *Journal of the American Medical Association* January 21, 1911; G. Klein, "The Tale of the Great Cuckoo Egg," *Nature* 400 (1999): 515.

**51** Jackalopes: D. G. McNeill, "How a Vaccine Search Ended in Triumph," *New York Times*, August 29, 2006.

**52** Chicken leukemia virus: W. F. Hughes, D. H. Watanabe, and H. Rubin, "The Development of a Chicken Flock Apparently Free of Leukosis Virus," *Avian Diseases* 7 (1963): 154–65; P. K. Vogt, "Avian Tumor Viruses," *Advances in Virus Research* 11 (1965): 293–385; T. Graf and H. Beug, "Avian Leukemia Viruses: Interaction with Their Target Cells in vivo and in vitro," *Biochemica et Biophysica Acta* 516 (1978): 269–99; M. J. Hayman, "Transforming Proteins of Avian Retroviruses," *Journal of General Virology* 52 (1981): 1–14; R. C. Gallo and F. Wong-Staal, "Retroviruses as Etiologic Agents of Some Animal and Human Leukemias and Lymphomas and as Tools for Elucidating the Molecular Mechanism of Leukemogenesis," *Blood*

60 (1982): 545–57; M. J. Hayman, "Avian Acute Leukemia Viruses," *Current Topics in Microbiology and Immunology* 103 (1983): 109–25; K. Blister and H. W. Jansen, "Oncogenes in Retroviruses and Cells: Biochemistry and Molecular Genetics," *Advances in Cancer Research* 47 (1986): 99–188; H. Beug, A. Bauer, H. Dolznig, et al., "Avian Erythryopoiesis and Erythroleukemia: Towards Understanding the Role of the Biomolecules Involved," *Biochemica et Biophysica Acta* 1288 (1996): M37–M47; A. M. Fadley, "Avian Retroviruses," *Veterinary Clinics of North America: Food and Animal Practice* 13 (1997): 71–85.

53 Rubin test: H. Rubin: "A Virus in Chick Embryos Which Induces Resistance in vitro to Infection with Rous Sarcoma Virus," *Proceedings of the National Academy of Science USA* 46 (1960): 1105–19," and "The Nature of a Virus-Induced Cellular Resistance to Rous Sarcoma Virus," *Virology* 13 (1961): 200–6.

53 Kimber Farms: P. Smith and Daniel, *Chicken Book*; "Museum of Local History Exhibits Kimber Memorabilia," *Tri-City Voice*, December 21, 2004; Fremont's economy: http://www.geography.berkeley.edu/ProjectsResources/CommunityProfiles/FremontProject/WebPage.

56 Yellow fever vaccine and cancer: T. D. Waters, P. S. Anderson, G. W. Beebe, and R. W. Miller, "Yellow Fever Vaccination, Avian Leukosis Virus, and Cancer Risk in Man," *Science* 177 (1972): 76–77.

## Coughs, Colds, Cancers, and Chickens

58 Marek's disease and vaccine: R. L. Witter, K. Nazerian, H. G. Purchase, and G. H. Burgoyne, "Isolation from Turkeys of a Cell-Associated Herpesvirus Antigenically Related to Marek's Disease Virus," *American Journal of Veterinary Research* 31 (1970): 525–38; M. R. Hilleman, "Marek's Disease Vaccine: Its Implications in Biology and Medicine," *Avian Diseases* 16 (1972): 191–99.

60 Hubbard Farms: "Tribute to Hubbard Farms on Their 75th Anniversary Celebration," http://thomas.loc.gov/cgi-bin/query/z?r104:S05SE6–1109; S. Aldag, "University of New Hampshire Will Honor Philanthropic Excellence on University Day," September 11, College of Life Sciences and Agriculture *Insight*, July 26, 2001, http://www.unh.edu/news/news_releases/2001/july%20/sa_20010726hubbard.html.

61 Penicillin: Lax, *Mold*.

61 Isaacs and Lindenmann: J. Lindenmann, "From Interference to Interferon: A Brief Historical Introduction," *Philosophical Transactions of the Royal Society of London* 299 (1982): 3–6; A. Isaacs and J. Lindenmann, "Virus Interference, I: The Interferon," *Journal of Interferon Research* 7 (1987): 429–38.

62 Hilleman's interferon studies: G. P. Lampson, A. A. Tytell, and M. R. Hilleman, "Purification and Characterization of Chick Embryo Interferon," *Proceedings of the Society for Experimental Biology and Medicine* 112 (1963): 468–78; G. Lampson, A. A. Tytell, M. M. Nemes, and M. R. Hilleman,

"Characterization of Chick Embryo Interferon Induced by a DNA Virus," *Proceedings of the Society for Experimental Biology and Medicine* 118 (1965): 441–48; G. P. Lampson, A. A. Tytell, M. M. Nemes, and M. R. Hilleman, "Multiple Molecular Species of Interferons of Mouse and of Rabbit Origin," *Proceedings of the Society for Experimental Biology and Medicine* 121 (1966): 377–84; G. P. Lampson, A. A. Tytell, A. K. Field, M. M. Nemes, and M. R. Hilleman, "Inducers of Interferon and Host Resistance, I: Double-Stranded RNA from Extracts of Penicillium funiculosum," *Proceedings of the National Academy of Sciences USA* 58 (1967): 782–89; A. K. Field, A. A. Tytell, G. P. Lampson, and M. R. Hilleman, "Inducers of Interferon and Host Resistance, II: Multi-Stranded Synthetic Polynucleotide Complexes," *Proceedings of the National Academy of Sciences USA* 58 (1967): 1004–10; A. A. Tytell, G. P. Lampson, A. K. Field, and M. R. Hilleman, "Inducers of Interferon and Host Resistance; III: Double-Stranded RNA from Reovirus Type 3 Virions (REO 3-RNA)," *Proceedings of the National Academy of Sciences USA* 58 (1967): 1719–22; A. K. Field, G. P. Lampson, A. A. Tytell, M. M. Nemes, and M. R. Hilleman, "Inducers of Interferon and Host Resistance; IV: Double-Stranded Replicative form RNA (MS2-RF-RNA) from E. coli Infected with MS2 Coliphage," *Proceedings of the National Academy of Sciences USA* 58 (1967): 2102–8; A. K. Field, A. A. Tytell, G. P. Lampson, and M. R. Hilleman, "Inducers of Interferon and Host Resistance; V: In vitro Studies," *Proceedings of the National Academy of Sciences USA* 61 (1968): 340–46; V. M. Larson, W. R. Clark, and M. R. Hilleman, "Influence of Synthetic (Poly I:C) and Viral Double-Stranded Ribonucleic Acids on Adenovirus 12 Oncogenesis in Hamsters," *Proceedings of the Society for Experimental Biology and Medicine* 131 (1969): 1002–11; V. M. Larson, W. R. Clark, G. E. Dagle, and M. R. Hilleman, "Influence of Synthetic Double-Stranded Ribonucleic Acid, Poly I:C, on Friend Leukemia Virus in Mice," *Proceedings of the Society for Experimental Biology and Medicine* 132 (1969): 602–7; M. M. Nemes, A. A. Tytell, G. P. Lampson, A. K. Field, and M. R. Hilleman, "Inducers of Interferon and Host Resistance; VI: Antiviral Efficacy of Poly I:C in Animal Models," *Proceedings of the Society for Experimental Biology and Medicine* 132 (1969): 776–83; M. M. Nemes, A. A. Tytell, G. P. Lampson, A. K. Field, and M. R. Hilleman, "Inducers of Interferon and Host Resistance; VII: Antiviral Efficacy of Double-Stranded RNA of Natural Origin," *Proceedings of the Society for Experimental Biology and Medicine* 132 (1969): 784–89; G. P. Lampson, A. A. Tytell, A. K. Field, M. M. Nemes, and M. R. Hilleman, "Influence of Polyamines on Induction of Interferon and Resistance to Viruses by Synthetic Polynucleotides," *Proceedings of the Society for Experimental Biology and Medicine* 132 (1969): 212–18; V. M. Larson, P. N. Panteleakis, and M. R. Hilleman, "Influence of Synthetic Double-Stranded Ribonucleic Acid (Poly I:C) on SV40 Viral Oncogenesis and Transplant Tumor in Hamsters," *Proceedings of the Society for Experimental Biology and Medicine* 133 (1970): 14–19; M. R. Hilleman, "Some Preclinical Studies in Animal Models with Double-Stranded RNA's," *Annals* of *the New York Academy of Sciences* 173 (1970): 623–28; G. P. Lampson, A. K. Field, A. A. Tytell, M. M. Nemes, and M. R. Hilleman, "Relationship of Molecular Size of $rI_u$:$rC_u$ (Poly I:C) to Induction of Interferon and Host Resistance," *Proceedings of the Society for Experimental Biology and Medicine* 135 (1970): 911–16; A. A. Tytell, G. P. Lampson, A. K. Field, M. M. Nemes, and M. R. Hilleman, "Influence of Size of Individual Homopolynucleotides on the Physical and Biological Properties of Complexed $rI_u$: $rC_u$ (Poly I:C)," *Proceedings of the Society for Experimental Biology and Medicine* 135 (1970): 917–21; C. L. Baugh, A. A. Tytell, and M. R. Hilleman, "In vitro Safety Assessment of Double-Stranded Polynucleotides Poly I:C and Mu–9," *Proceedings of the Society for Experimental Biology and Medicine* 137 (1971): 1194–98; A. K. Field, A. A. Tytell, G. P. Lampson, and M. R. Hilleman, "Antigenicity of Double-Stranded Ribonucleic Acids Including Poly I: C," *Proceedings of the Society for Experimental Biology and Medicine* 139 (1972): 1113–19; A. K. Field, A. A. Tytell, G. P. Lampson, and M. R. Hilleman, "Cellular Control of Interferon Production and Release after Treatment with Poly I:C Inducer," *Proceedings of the Society for Experimental Biology and Medicine* 140 (1972): 710–14; A. K. Field, G. P. Lampson, A. A. Tytell, and M. R. Hilleman, "Demonstration of Double-Stranded Ribonucleic Acid in Concentrates of RNA Viruses," *Proceedings of the Society for Experimental Biology and Medicine* 141 (1972): 440–44; G. P. Lampson, M. M. Nemes, A. K. Field, A. A. Tytell, and M. R. Hilleman, "The Effect of Altering the Size of Poly C on the Toxicity and Antigenicity of Poly I:C," *Proceedings of the Society for Experimental Biology and Medicine* 141 (1972): 1068–72; B. Mendlowski, A. K. Field, A. A. Tytell, and M. R. Hilleman, "Safety Assessment of Poly I:C in NZB/NZW Mice," *Proceedings of the Society for Experimental Biology and Medicine* 148 (1975): 476–83; G. P. Lampson, A. K. Field, A. A. Tytell, and M. R. Hilleman, "Poly I:C/Poly-L-Lysine: Potent Inducer of Interferons in Primates," *Journal of Interferon Research* 1 (1981): 539–49.

63 Japanese Encephalitis Virus: Plotkin and Orenstein, *Vaccines*.

65 History of the common cold: Williams, *Virus Hunters*; R. B. Couch, "Rhinoviruses," in *Fields Virology*, ed. D. M. Knipe and P. H. Howley, 4th ed. (Philadelphia: Lippincott Williams and Wilkins, 2001).

66 Winston Price studies: Williams, *Virus Hunters*; "Text of Dr. Price's Report on Development of Cold Vaccine," *New York Times*, September 19, 1957; "Vaccine for a Type of Cold Reported Success in Tests," *New York Times*, September 19, 1957; "New Study Set on Cold Vaccines," *New York Times*, September 20, 1957; "Vaccine Hailed in City," *New York Times*, September 20, 1957; D. G. Cooley, "Fighting the Common Cold," *New York Times*, September 22, 1957; "Visit to a Common Cold Laboratory," *New York Times*, November 3, 1957.

67 Hilleman's studies of the common cold: Williams, *Virus Hunters*; V. V. Hamparian, A. Ketler, and M. R. Hilleman, "Recovery of New Viruses (Coryzavirus) from Cases of Common Cold in Human Adults," *Proceedings of the Society for Experimental Biology and Medicine* 108 (1961): 444–53;

C. M. Reilly, S. M. Hoch, J. Stokes Jr., L. McClelland, V. V. Hamparian, A. Ketler, and M. R. Hilleman, "Clinical and Laboratory Findings in Cases of Respiratory Illness Caused by Coryzaviruses," *Annals of Internal Medicine* 57 (1962): 515–25; A. Ketler, V. V. Hamparian, and M. R. Hilleman, "Characterization and Classification of ECHO 28-Rhinvovirus-Coryzavirus Agents," *Proceedings of the Society for Experimental Biology and Medicine* 110 (1962): 821–31; C. M. Reilly, J. Stokes Jr., V. V. Hamparian, and M. R. Hilleman, "ECHO Virus, Type 25, in Respiratory Illness," *Journal of Pediatrics* 62 (1963): 536–39; V. V. Hamparian, M. R. Hilleman, and A. Ketler, "Contributions to Characterization and Classification of Animal Viruses," *Proceedings of the Society for Experimental Biology and Medicine* 112 (1963): 1040–50; V. V. Hamparian, A. Ketler, and M. R. Hilleman, "The ECHO 28-Rhinovirus-Coryzavirus (ERC) Group of Viruses," *American Review of Respiratory Diseases* 88 (1963): 269–73; V. V. Hamparian, M. B. Leagus, M. R. Hilleman, and J. Stokes Jr., "Epidemiologic Investigations of Rhinovirus Infections," *Proceedings of the Society for Experimental Biology and Medicine* 117 (1964): 469–76; V. V. Hamparian, M. B. Leagus, and M. R. Hilleman, "Additional Rhinovirus Serotypes," *Proceedings of the Society for Experimental Biology and Medicine* 116 (1964): 976–84; C. C. Mascoli, M. B. Leagus, R. E. Weibel, H. Reinhart, and M. R. Hilleman, "Attempt at Immunization by Oral Feeding of Live Rhinovirus in Enteric-Coated Capsules," *Proceedings of the Society for Experimental Biology and Medicine* 121 (1966): 1264–68; C. C. Mascoli, M. B. Leagus, M. R. Hilleman, R. E. Weibel, and J. Stokes Jr., "Rhinovirus Infection in Nursery and Kindergarten Children: New Rhinovirus Serotype 54," *Proceedings of the Society for Experimental Biology and Medicine* 124 (1967): 845–50; M. R. Hilleman, C. M. Reilly, J. Stokes Jr., and V. V. Hamparian, "Clinical-Epidemiologic Findings in Coryzavirus Infections," *American Review of Respiratory Diseases* 88 (1963): 274–76; M. R. Hilleman, "Present Knowledge of the Rhinovirus Group of Viruses," *Ergebnisse der Mikrobiologie Immunitatsforschun und Experimentellen Therapie* 41 (1967): 1–22.

**The Monster Maker**

Stanley Plotkin was interviewed on February 15, 2005; Leonard Hayflick on November 6 and 7, 2006.

69 Rubella disease: Plotkin and Orenstein, *Vaccines*.

70 Gregg: N. M. Gregg, "Congenital Cataract Following German Measles in the Mother," *Transactions of the Ophthalmologic Society of Australia* 3 (1941): 35–46; "Sir Norman McAlister Gregg, 1892–1966," *American Journal of Ophthalmology* 63 (1967): 180–81.

71 Rubella epidemic, 1963–1964: Plotkin and Orenstein, *Vaccines*.

71 Vietnam war deaths: National Archives: http://www.archives.gov/research/vietnam-war/casualty-statistics.html.

72 Hilleman rubella vaccine studies: E. B. Buynak, M. R. Hilleman, R. E. Weibel, and J. Stokes Jr., "Live Attenuated Rubella Virus Vaccines Prepared

in Duck Embryo Cell Culture," *Journal of the American Medical Association* 204 (1968): 195–200; R. E. Weibel, J. Stokes Jr., E. B. Buynak, J. E. Whitman, M. B. Leagus, and M. R. Hilleman, "Live Attenuated Rubella Virus Vaccines Prepared in Duck Embryo Cell Culture, II: Clinical Tests in Families in an Institution," *Journal of the American Medical Association* 205 (1968): 554–58; M. R. Hilleman, E. B. Buynak, R. E. Weibel, and J. Stokes Jr., "Live Attenuated Rubella Vaccine," *New England Journal of Medicine* 279 (1968): 300–303; M. R. Hilleman, "Toward Prophylaxis of Prenatal Infection by Viruses," *Obstetrics and Gynecology* 33 (1969): 461–69; M. R. Hilleman, E. B. Buynak, J. E. Whitman Jr., R. E. Weibel, and J. Stokes Jr., "Summary Report on Rubella Virus Vaccines Prepared in Duck Embryo Cell Culture," International Symposium on Rubella Vaccines, *Symposium Series on Immunobiological Standards* 11 (1969): 349–56; M. R. Hilleman, E. B. Buynak, J. E. Whitman Jr., R. E. Weibel, and J. Stokes Jr., "Live Attenuated Rubella Virus Vaccines," *American Journal of Diseases in Children* 118 (1969): 166–71; J. Stokes Jr., R. E. Weibel, E. B. Buynak, and M. R. Hilleman, "Clinical-Laboratory Findings in Adult Women Given HPV–77 Rubella Vaccine, International Symposium on Rubella Vaccines, *Symposium Series Immunobiological Standards* 11 (1969): 415–22; R. E. Weibel, J. Stokes Jr., E. B. Buynak, and M. R. Hilleman, "Rubella Vaccination in Adult Females," *New England Journal of Medicine* 280 (1969): 682–85; J. Stokes Jr., R. E. Weibel, E. B. Buynak, and M. R. Hilleman, "Protective Efficacy of Duck Embryo Rubella Vaccines," *Pediatrics* 44 (1969): 217–24; R. E. Weibel, J. Stokes Jr., E. B. Buynak, and M. R. Hilleman, "Live Rubella Vaccines in Adults and Children," *American Journal of Diseases in Children* 118 (1969): 226–29; E. B. Buynak, V. M. Larson, W. J. McAleer, C. C. Mascoli, and M. R. Hilleman, "Preparation and Testing of Duck Embryo Cell Culture Rubella Vaccine," *American Journal of Diseases of Children* 118 (1969): 347–54; T. A. Swartz, W. Klingberg, R. A. Goldwasser, M. A. Klingberg, N. Goldblum, and M. R. Hilleman, "Clinical Manifestations, According to Age, among Females Given HPV–77 Duck Rubella Vaccine," *American Journal of Epidemiology* 94 (1971): 246–51; R. E. Weibel, J. Stokes Jr., E. R. Buynak, and M. R. Hilleman, "Influence of Age on Clinical Response to HPV–77 Duck Rubella Vaccine," *Journal of the American Medical Association* 222 (1972): 805–7; V. M. Villarejos, J. A. Arguedas, G. C. Hernandez, E. B. Buynak, and M. R. Hilleman, "Clinical Laboratory Evaluation of Rubella Vaccine Given to Postpartum Women without Pregnancy Preventive," *Obstetrics and Gynecology* 42 (1973): 689–95; R. E. Weibel, V. M. Villarejos, E. B. Klein, E. B. Buynak, A. A. McLean, and M. R. Hilleman, "Clinical and Laboratory Studies of Live Attenuated RA27/3 and HPV77-DE Rubella Virus Vaccines," *Proceedings of the Society for Experimental Biology and Medicine* 165 (1980): 44–49.

72 Mary Lasker: Mary Lasker Papers, 1940–1993. http://www.columbia.edu/cu/libraries/indiv/rare/guides/Lasker/main.html; Congressional Gold Medal Recipients, Mary Woodard Lasker (1900–1994), http://www.congressionalgoldmedal.com/MaryLasker.htm; Mary Lasker's activism in

the birth control movement, http://ourworld.compuserve.com/homepages/CarolASThompson/birthcon.html.

75 Max Tishler: Max Tishler: Biographical memoirs, http://www.nap.edu/readingroom/books/biomems/mtishler.html; Max Tishler: Professor of Chemistry, 1970–1987, http://wesleyan.edu/chem/Leermakers/max_tishler.html; National Inventors Hall of Fame: Max Tishler, http://www.invent.org/hall_of_fame/145.html.

76 Roy Vagelos quote: Symposium in honor of Maurice R. Hilleman, American Philosophical Society, January 26, 2005.

77 Meyer and Parkman rubella vaccine studies: P. D. Parkman, H. M. Meyer, R. L. Kirchstein, and H. E. Hopps, "Attenuated Rubella Virus, I: Development and Laboratory Characterization," *New England Journal of Medicine* 275 (1966): 569–74; H. M. Meyer, P. D. Parkman, and T. C. Panos, "Attenuated Rubella Virus, II: Production of an Experimental Live-Virus Vaccine and Clinical Trial," *New England Journal of Medicine* 275 (1966): 575–80.

79 Wistar Institute: The Wistar Institute, http://www.wistar.upenn.edu/about_wistar/history.html; Caspar Wistar Papers, American Philosophical Society, http://www.amphilsoc.org/library/mole/w/wistar.htm.

80 Koprowski and Sabin polio vaccines: Oshinsky, *Polio*; J. Smith, *Patenting the Sun*; Carter, *Breakthrough*.

81 Hippocratic Oath: Hippocratic Oath, classical version, http://www.pbs.org/wgbh/nova/doctors/oath_classical.html; Hippocratic Oath, modern version, http://www.pbs.org/wgbh/nova/doctors/oath_modern.html; "The Hippocratic Oath Today: Meaningless Relic or Invaluable Guide," http://www.pbs.org/wgbh/nova/doctors/oath_today.html.

83 Hayflick studies: Hall, *Merchants of Immortality*; J. W. Shay, and W. E. Wright, "Hayflick, His Limit, and Cellular Ageing," *Nature* 1 (2000): 72–76; L. Hayflick, "The Limited in vitro Lifetime of Human Diploid Cell Strains," *Experimental Cell Research* 37 (1965): 614–36.

85 Telomeres and aging: M. A. Blasco, "Telomeres and Human Disease: Aging, Cancer and Beyond," *Nature Reviews Genetics* 6 (2005): 611–12; S, E, Artandi, "Telomeres, Telomerase, and Human Disease," *New England Journal of Medicine* 355 (2006): 1195–97.

85 Anaconda movie: *Anacondas: The Hunt for the Blood Orchid*, Sony Pictures, 2004.

86 Plotkin rubella vaccine studies: S. A. Plotkin, D. Cornfeld, and T. H. Ingalls, "Studies of Immunization with Living Rubella Virus," *American Journal of Diseases of Children* 110 (1965): 381–89; S. A. Plotkin, J. Farquhar, M. Katz, and T. H. Ingalls, "A New Attenuated Rubella Virus Grown in Human Fibroblasts: Evidence for Reduced Nasopharyngeal Excretion," *American Journal of Epidemiology* 86 (1967): 468–77; S. Saidi and K. Naficy, "Subcutaneous and Intranasal Administration of RA27/3 Rubella Vaccine," *American Journal of Diseases of Children* 118 (1969): 209–12; S. A. Plotkin, J. D. Farquhar, and M. Katz, "Attenuation of RA27/3 Rubella Virus in WI–38 Human Diploid Cells," *American Journal of Diseases of Children* 118

(1969): 178–85; D. M. Horstmann, H. Liebhaber, G. L. Le Bouvier, D. A. Rosenberg, and S. B. Halstead, "Rubella: Reinfection of Vaccinated and Naturally Immune Persons Exposed in an Epidemic," *New England Journal of Medicine* 283 (1970): 771–78; A. Fogel, A. Moshkowitz, L. Rannon, and C. B. Gerichter, "Comparative Trials of RA27/3 and Cendehill Rubella Vaccines in Adult and Adolescent Females," *American Journal of Epidemiology* 93 (1971): 392–98; H. Liebhaber, T. H. Ingalls, G. L. Le Bouvier, and D. M. Horstmann, "Vaccination with RA27/3 Rubella Vaccine," *American Journal of Diseases of Children* 123 (1972): 133–36; S. L. Spruance, L. E. Klock, A. Bailey, J. R. Ward, and C. B. Smith, "Recurrent Joint Symptoms in Children Vaccinated with HPV77DK12 Rubella Vaccine," *Journal of Pediatrics* 80 (1972): 413–17; S. A. Plotkin, J. D. Farquhar, and P. L. Ogra, "Immunologic Properties of RA27/3 Rubella Virus Vaccine: A Comparison with Strains Presently Licensed in the United States," *Journal of the American Medical Association* 225 (1973): 585–90; S. L. Spruance, R. Metcalf, C. B. Smith, M. M. Griffiths, and J. R. Ward, "Chronic Arthropathy Associated with Rubella Vaccination," *Arthritis and Rheumatism* 20 (1977): 741–77; B. F. Polk, J. F. Modlin, J. A. White, and P. C. DeGirolami, "A Controlled Comparison of Joint Reactions among Women Receiving One of Two Rubella Vaccines," *American Journal of Epidemiology* 115 (1982): 19–25; H. L. Nakhasi, D. Thomas, D. Zheng, and T.-Y. Liu, "Nucleotide Sequence of Capsid, E2, and E1 Protein Genes of Rubella Virus Vaccine Strain RA27/3," *Nucleic Acids Research* 17 (1989): 4393–94.

88 Studies of rubella vaccine in pregnant women: Plotkin and Orenstein, *Vaccines*; W. F. Fleet, E. W. Benz, D. T. Karzon, L. B. Lefkowitz, and K. L. Herrmann, "Fetal Consequences of Maternal Rubella Immunization," *Journal of the American Medical Association* 227 (1974): 621–27.

88 Gerberding announcement: "CDC Announces Rubella, Once a Major Cause of Birth Defects, Is No Longer a Health Threat in the U.S.," telebriefing transcript, March 21, 2005.

**Political Science**

Phil Provost was interviewed on May 15, 2006; Stanley Plotkin on February 15, 2005, and September 15, 2006; Keerti Shah on May 22, 2006, and Leonard Hayflick on November 6 and 7, 2006.

90 Debi Vinnedge and the Catholic Church: Catholic Exchange, "Vaccines and Abortion: What's the Right Choice," http://www.catholicexchange.com/vm/index.asp?art_id=31229; American Life League, "Activism: Vaccines and the Catholic Doctrine," http://www.vaclib.org/basic/activism.htm; C. Glatz, "Vatican Says Refusing Vaccines Must Be Weighed against Health Threats," Catholic News Service, http://www.catholicnews.com/data/stories/cns/0504240.htm; "Vatican Statement on Vaccines Derived from Aborted Human Fetuses," Pontifica Academia Pro Vita, http://www.immunize.org/concerns/vaticandocument.htm; Debi Vinnedge and Children of God for Life, http://www.prolifepac.com/ html/who19vinnedge.htm; "Vaccines and

Fetuses," The Millennium Project, http://www.ratbags.com/rsoles/vaxliars/foeti.htm; "Vaccines Originating in Abortion," *Ethics and Medics* 24 (1999): 3–4; "Moral Reflection on Vaccines Prepared from Cells Derived from Aborted Human Foetuses," Pontifical Academy for Life, Congregation for the Doctrine of Faith, http://www.consciencelaws.org/Conscience-Policies-Papers/PPPCatholic03.html; D. Vinnedge: "Responding to the Call: Is Anyone Listening?" http://www.cogforlife.org/responding.htm, and "Vaccines from Abortion: The Truth," http://www.all.org/celebrate_life/c10107c.htm; National Network for Immunization Information, "Vaccine Components: Human Fetal Links with Some Vaccines," http://www.immunizationinfo.org/vaccine_ components_ detail.cfv?id=32.

95 Hilleman SV40 studies: B. H. Sweet and M. R. Hilleman, "Detection of a 'Non-Detectable' Simian Virus (Vacuolating Agent) Present in Rhesus and Cynomolgus Monkey-Kidney Cells Culture Material: A Preliminary Report," Second International Conference on Live Poliovirus Vaccines, Pan American Health Organization and World Health Organization, Washington, D.C., June 1960; B. H. Sweet and M. R. Hilleman, "The Vacuolating Virus, SV40," *Proceedings of the Society for Experimental Biology and Medicine* 105 (1960): 420–27; A. J. Girardi, B. H. Sweet, V. B. Slotnick, and M. R. Hilleman, "Development of Tumors in Hamsters Inoculated in the Neonatal Period with Vacuolating Virus, SV40," *Proceedings of the Society for Experimental Biology and Medicine* 109 (1962): 649–60; A. J. Girardi, B. H. Sweet, and M. R. Hilleman, "Factors Influencing Tumor Induction in Hamsters by Vacuolating Virus, SV40," *Proceedings of the Society for Experimental Biology and Medicine* 112 (1963): 662–67; A. J. Girardi, and M. R. Hilleman, "Host-Virus Relationships in Hamsters Inoculated with SV40 Virus during the Neonatal Period," *Proceedings of the Society for Experimental Biology and Medicine* 116 (1964): 723–28; H. Goldner, A. J. Girardi, V. M. Larson, and M. R. Hilleman, "Interruption of SV40 Virus Tumorigenesis Using Irradiated Homologous Tumor Antigen," *Proceedings of the Society for Experimental Biology and Medicine* 117 (1964): 851–57; H. Goldner, A. J. Girardi, and M. R. Hilleman, "Enhancement in Hamsters of Virus Oncogenesis Attending Vaccination Procedures," *Virology* 27 (1965): 225–27; J. H. Coggin, V. M. Larson, and M. R. Hilleman, "Prevention of SV40 Virus Tumorigenesis by Irradiated, Disrupted and Iododeoxyuridine-Treated Tumor Cell Antigens," *Proceedings of the Society for Experimental Biology and Medicine* 124 (1967): 774–84; V. M. Larson, W. G. Raupp, and M. R. Hilleman, "Prevention of SV40 Virus Tumorigenesis in Newborn Hamsters by Maternal Immunization," *Proceedings of the Society for Experimental Biology and Medicine* 126 (1967): 674–77; V. M. Larson, W. R. Clark, and M. R. Hilleman, "Cryosurgical Treatment of Primary SV40 Viral and Adenovirus 7 Transplant Tumors in Hamsters," *Proceedings of the Society for Experimental Biology and Medicine* 128 (1968): 983–88; P. N. Panteleakis, V. M. Larson, E. S. Glenn, and M. R. Hilleman, "Prevention of Viral and Transplant Tumors in Hamsters Employing Killed and Fragmented Homologous Tumor

Cell Vaccines," *Proceedings of the Society for Experimental Biology and Medicine* 129 (1968): 50–57; V. M. Larson, W. R. Clark, and M. R. Hilleman, "Comparative Studies of SV40 and Adenovirus Oncogenesis in Random Bred and Inbred Hamsters," *Proceedings of the Society for Experimental Biology and Medicine* 137 (1971): 607–13.

97 SV40 and cancer: M. D. Innis, "Oncogenesis and Poliomyelitis Vaccine," *Nature* 219 (1968): 972–73; J. F. Fraumeni Jr., C. R. Stark, E. Gold, and M. L. Lepow, "Simian Virus 40 in Polio Vaccine: Follow-Up of Newborn Recipients," *Science* 167 (1970): 59–60; K. V. Shah, H. L. Ozer, H. S. Pond, et al., "SV40 Neutralizing Antibodies in Sera of U.S. Residents without History of Polio Immunization," *Nature* 231 (1971): 448–49; K. Shah and N. Nathanson, "Human Exposure to SV40: Review and Comment," *American Journal of Epidemiology* 103 (1976): 1–12; E. A. Mortimer, M. L. Lepow, E. Gold, et al., "Long-Term Follow-Up of Persons Inadvertently Inoculated with SV40 as Neonates," *New England Journal of Medicine* 305 (1981): 1517–18; B. Kuska, "SV40: Working the Bugs out of the Polio Vaccine," *Journal of the National Cancer Institute* 89 (1997): 283–84; M. Carbone, P. Rizzo, and H. I. Pass, "Simian Virus 40, Polio Vaccines and Human Tumors: A Review of Recent Developments," *Oncogene* 15 (1997): 1877–88; L. Hayflick, "SV40 and Human Cancer," *Science* 276 (1997): 337–38; A. Procopio, R. Marinacci, M. R. Marinetti, et al., "SV40 Expression in Human Neoplastic and Non-Neoplastic Tissues: Perspectives on Diagnosis, Prognosis and Therapy of Human Malignant Melanoma," *Development of Biological Standards* 94 (1998): 361–67; P. Olin and J. Giesecke, "Potential Exposure to SV40 in Polio Vaccines Used in Sweden during 1957: No Impact on Cancer Incidence Rates 1960 to 1993," *Development of Biological Standards* 94 (1998): 227–33; H. D. Strickler and J. J. Goedert, "Exposure to SV40-Contaminated Poliovirus Vaccine and the Risk of Cancer: A Review of the Epidemiologic Evidence," *Development of Biological Standards* 94 (1998): 235–44; S. C. Stenton, "Simian Virus 40 and Human Malignancy," *British Medical Journal* 316 (1998): 877; H. D. Strickler, P. S. Rosenberg, S. S. Devesa, et al., "Contamination of Poliovirus Vaccines with Simian Virus 40 (1955–1963) and Subsequent Cancer Rates," *Journal of the American Medical Association* 279 (1998): 292–95; H. D. Strickler, P. S. Rosenberg, S. S. Devesa, et al., "Contamination of Poliovirus Vaccine with SV40 and the Incidence of Medulloblastoma," *Medical and Pediatric Oncology* 32 (1999): 77–78; S. G. Fisher, L. Weber, and M. Carbone, "Cancer Risk Associated with Simian Virus 40 Contaminated Polio Vaccine," *Anticancer Research* 19 (1999): 2173–80; D. Sangar, P. A. Pipkin, D. J. Wood, and P. D. Minor, "Examination of Poliovirus Vaccine Preparations for SV40 Sequences," *Biologicals* 27 (1999): 1–10; M. R. Goldman and M. J. Brock, "Contaminated Polio Vaccines: Will the Next Shot Be Fired in the Courtroom?" *Journal of Legal Medicine* 20 (1999): 223–49; J. S. Butel, "Simian Virus 40, Poliovirus Vaccines, and Human Cancer: Research Progress versus Media and Public Interests," *Bulletin of the World Health Organization* 78 (2000): 195–98; H. Ohgaki, H. Huang, M. Haltia, et al.,

"More about Cell and Molecular Biology of Simian Virus 40: Implications for Human Infection and Disease," *Journal of the National Cancer Institute* 92 (2000): 495–96; C. Carroll-Pankhurst, E. A. Engels, H. D. Strickler, et al., "Thirty-five Year Mortality Following Receipt of SV40-Contaminated Polio Vaccine during the Neonatal Period," *British Journal of Cancer* 85 (2001): 1295–97; D. Ferber, "Creeping Consensus on SV40 and Polio Vaccine," *Science* 298 (2002): 725–27; M. Carbone, H. I. Pass, L. Miele, and M. Bocchetta, "New Developments about the Association of SV40 with Human Mesothelioma," *Oncogene* 22 (2003): 5173–80; P. Minor, P. Pipkin, Z. Jarzebek, and W. Knowles, "Studies of Neutralizing Antibodies to SV40 in Human Sera," *Journal of Medical Virology* 70 (2003): 490–95; E. A. Engels, L. H. Rodman, M. Frisch, et al., "Childhood Exposure to Simian Virus 40-Contaminated Poliovirus Vaccine and Risk of AIDS-Associated Non-Hodgkin's Lymphoma," *International Journal of Cancer* 106 (2003): 283–87; R. A. Vilchez, A. S. Arrington, and J. S. Butel, "Cancer Incidence in Denmark Following Exposure to Poliovirus Vaccine Contaminated with Simian Virus 40," *Journal of the National Cancer Institute* 95 (2003): 1249; H. D. Strickler, J. J. Goedert, S. S. Devesa, et al., "Trends in U.S. Pleural Mesothelioma Incidence Rates Following Simian Virus 40 Contamination of Early Poliovirus Vaccines," *Journal of the National Cancer Institute* 95 (2003): 38–45; E. A. Engels, H. A. Katki, N. M. Nielson, et al., "Cancer Incidence in Denmark Following Exposure to Poliovirus Vaccine Contaminated with Simian Virus 40," *Journal of the National Cancer Institute* 95 (2003): 532–39; F. Mayall, K. Barratt, and J. Shanks, "The Detection of Simian Virus 40 in Mesotheliomas from New Zealand and England Using Real Time FRET Probe PCR Protocols," *Journal of Clinical Pathology* 56 (2003): 728–30; M. Carbone, and M. A. Rkzanek, "Pathogenesis of Malignant Mesothelioma," *Clinical Lung Cancer* 5 (2004): S46–S50; G. Barbanti-Brodano, S. Sabbioni, F. Martini, et al., "Simian Virus 40 Infection in Humans and Association with Human Diseases: Results and Hypotheses," *Virology* 318 (2004): 1–9; K. V. Shah, "Simian Virus 40 and Human Disease," *Journal of Infectious Diseases* 190 (2004): 2061–64; M. Jin, H. Sawa, T. Suzuki, et al., "Investigation of Simian Virus 40 Large T Antigen in 18 Autopsied Malignant Mesothelioma Patients in Japan," *Journal of Medical Virology* 74 (2004): 668–76; D. E. M. Rollison, W. F. Page, H. Crawford, et al., "Case-Control Study of Cancer among U.S. Army Veterans Exposed to Simian Virus 40-Contaminated Adenovirus Vaccine," *American Journal of Epidemiology* 160 (2004): 317–24; E. A. Engels, J. Chen, R. P. Viscidi, et al., "Poliovirus Vaccination during Pregnancy, Maternal Seroconversion to Simian Virus 40, and Risk of Childhood Cancer," *American Journal of Epidemiology* 160 (2004): 306–16.

**99** Polio vaccine as the source of AIDS: E. Hooper, *The River: A Journey to the Source of HIV and AIDS* (Boston: Little, Brown, 1999); S. A. Plotkin and H. Koprowski, "No Evidence to Link Polio Vaccine with HIV," *Nature* 407 (2000): 941; S. A. Plotkin, D. E. Teuwen, A. Prinzie, and J. Desmyter, "Postscript Relating to New Allegations Made by Edward Hooper at the Royal

Society Discussion Meeting on 11 September 2000," *Philosophical Transactions of the Royal Society of London* 356 (2001): 825–29; S. A. Plotkin, "Untruths and Consequences: The False Hypothesis Linking CHAT Type 1 Polio Vaccination to the Origin of Human Immunodeficiency Virus," *Philosophical Transactions of the Royal Society of London* 356 (2001): 815–23; S. A. Plotkin, "Chimpanzees and Journalists," *Vaccine* 22 (2004): 1829–30.

**100** Source of HIV: L. Neergaard, "Scientists Trace AIDS Origin to Wild Chimps: Gene Tests Match Virus to Primates in Cameroon to First Known Human Case," Associated Press, May 25, 2006; L. Roberts, "Polio Eradication: Is It Time to Give Up?" *Science* 312 (2006): 832–35.

**100** Wistar rabies vaccine studies: T. J. Wiktor, F. Sokol, E. Kuwert, and H. Koprowski, "Immunogenicity of Concentrated and Purified Rabies Vaccine of Tissue Culture Origin," *Proceedings of the Society of Experimental Biology and Medicine* 131 (1969): 799–805; T. J. Wiktor, S. A. Plotkin, and D. W. Grella, "Human Cell Culture Rabies Vaccine," *Journal of the American Medical Association* 224 (1973): 1170–71.

**102** Chickenpox disease: Plotkin and Orenstein, *Vaccines*.

**102** Weller and chickenpox virus: Weller, *Growing Pathogens*.

**102** Takahashi and chickenpox vaccine: M. Takahashi, T. Otsuka, Y. Okuno, et al., "Live Vaccine Used to Prevent the Spread of Varicella in Children in Hospital," *Lancet* 2 (1974): 1288–90; M. Takahashi, Y. Okuno, T. Otsuka, et al., "Development of a Live Attenuated Varicella Vaccine," *Biken Journal* 18 (1975): 25–33.

**103** Hilleman and chickenpox vaccine: R. E. Weibel, B. J. Neff, B. J. Kuter, et al., "Live Attenuated Varicella Virus Vaccine: Efficacy Trial in Healthy Children," *New England Journal of Medicine* 310 (1984): 1409–15.

**103** Hepatitis A virus outbreak at Chi-Chi's: "Officials Link Chi-Chi's Hepatitis Outbreak to Green Onions," *USA Today*, November 21, 2003; S. Waite, "Thousands at Risk of Hepatitis," *Beaver County Times*, November 5, 2003; C. Snowbeck, writing in *Pittsburgh Post-Gazette*: "Hepatitis Outbreak in Beaver County Reaches 130," November 7, 2003; "Hepatitis Outbreak Claims First Fatality," November 8, 2003; "240 Cases of Hepatitis Listed in Beaver," November 11, 2003; "Hepatitis Outbreak Reaches 300," November 11, 2003; "How Hepatitis A Was Spread Remains a Mystery in Beaver County," November 12, 2003; "Beaver County Hepatitis Probe Changes Focus," November 13, 2003; "Second Death in Hepatitis Outbreak," November 14, 2003; "Hepatitis Probe Following Pattern," November 16, 2003; "Investigation Lets Chi-Chi's Staff off Hook," November 19, 2003; "FDA Stops Green Onions from 3 Mexican Suppliers," November 21, 2003; "Mexico Closes 4 Green Onion Exporters," November 25, 2003; "'Smoking Gun' in Outbreak Will Be Hard to Find," November 27, 2003; "How Going Out For a 'Decent Meal' Led to Transplant for Beaver Man," December 2, 2003; "Hepatitis Cases Rise to 635 in Beaver County," December 4, 2003; C. Sheehan, "PA. Hepatitis A Outbreak Kills 3rd Victim," Associated Press, November 14, 2003; L. Polgreen, "Community Is Reeling from Hepatitis Outbreak," *New York Times*, November 17, 2003; B. Bauder, "Hepatitis Cause Eludes Officials," *Beaver County Times*, November 18, 2003; B. Batz, "Hepatitis News Affecting Beaver County Residents in Different Ways," *Pittsburgh Post-Gazette*, November 18, 2003; A. Manning and E. Weise, "Hepatitis A Outbreak Tied to Imported Food," *USA Today*, November 19, 2003; "U.S. Bars Mexican Onions Due to Hepatitis Outbreak," Reuters, November 19, 2003; "Onions Blamed for Deadly Virus," CBS/AP, November 21, 2003; "Toll of Hepatitis A Outbreak Climbing," *Pittsburgh News-Leader*, November 21, 2003; J. Mandak, "Chi-Chi's Exec Calls Restaurants Safe," Associated Press, November 22, 2003; K. Roebuck, "Hepatitis Victims Describe Ordeals," *Tribune-Review*, April 25, 2004; C. Wheeler, T. M. Vogt, G. L. Armstrong, et al., "An Outbreak of Hepatitis A Associated with Green Onions, *New England Journal of Medicine* 353 (2005): 890–97.

105 Shanghai outbreak: G. Yao, "Clinical Spectrum and Natural History of Viral Hepatitis A in a 1988 Shanghai Epidemic," in *Viral Hepatitis and Liver Diseases* (Baltimore: Williams and Wilkins, 1991), 76–77.

105 Deinhardt studies: R. Deinhardt, A. W. Holmes, R. B. Capps, and H. Popper, "Studies on the Transmission of Human Viral Hepatitis to Marmoset Monkeys, I: Transmission of Disease, Serial Passages, and Description of Liver Lesions," *Journal of Experimental Medicine* 125 (1967): 673–88.

106 Deinhardt obituary: M. R. Hilleman, "A Tribute to Dr. Friedrich W. Deinhardt MD: 1926–1992," *Journal of Hepatology* 18 (1993): S2–S4.

107 Hilleman hepatitis A vaccine studies: C. C. Mascoli, O. L. Ittensohn, V. M. Villarejos, J. A. Arguedas, P. J. Provost, and M. R. Hilleman, "Recovery of Hepatitis Agents in the Marmoset from Human Cases Occurring in Costa Rica," *Proceedings of the Society for Experimental Biology and Medicine* 142 (1973): 276–82; P. J. Provost, O. L. Ittensohn, V. M. Villarejos, J. A. Arguedas, and M. R. Hilleman, "Etiologic Relationship of Marmoset-Propagated CR326 Hepatitis A Virus to Hepatitis in Man," *Proceedings of the Society for Experimental Biology and Medicine* 142 (1973): 1257–67; P. J. Provost, O. L. Ittensohn, V. M. Villarejos, and M. R. Hilleman, "A Specific Complement-Fixation Test for Human Hepatitis A Employing CR326 Virus Antigen: Diagnosis and Epidemiology," *Proceedings of the Society for Experimental Biology and Medicine* 148 (1975): 962–69; P. J. Provost, B. S. Wolanski, W. J. Miller, O. L. Ittensohn, W. J. McAleer, and M. R. Hilleman, "Physical, Chemical and Morphologic Dimensions of Human Hepatitis A Virus Strain CR326," *Proceedings of the Society for Experimental Biology and Medicine* 148 (1975): 532–39; M. R. Hilleman, P. J. Provost, W. J. Miller, et al., "Immune Adherence and Complement-Fixation Tests for Human Hepatitis A: Diagnostic and Epidemiologic Investigations," *Development of Biological Standards* 30 (1975): 383–89; M. R. Hilleman, P. J. Provost, B. S. Wolanski, et al., "Characterization of CR326 Human Hepatitis A Virus, a Probable Enterovirus," *Development of Biological Standards* 30 (1975): 418–24; W. J. Miller, P. J. Provost, W. J. McAleer, O. L. Ittensohn, V. M. Vil-

larejos, and M. R. Hilleman, "Specific Immune Adherence Assay for Human Hepatitis A Antibody: Application to Diagnostic and Epidemiologic Investigations," *Proceedings of the Society for Experimental Biology and Medicine* 149 (1975): 254–61; P. J. Provost, B. S. Wolanski, W. J. Miller, O. L. Ittensohn, W. J. McAleer, and M. R. Hilleman, "Biophysical and Biochemical Properties of CR326 Human Hepatitis A Virus," *American Journal of Medical Sciences* 270 (1975): 87–91; M. R. Hilleman, P. J. Provost, W. J. Miller, et al., "Development and Utilization of Complement-Fixation and Immune Adherence Tests for Human Hepatitis A Virus and Antibody," *American Journal of Medical Sciences* 270 (1975): 93–98; V. M. Villarejos, A. Gutierrez-Diermissen, K. Anderson-Visona, A. Rodriguez-Aragones, P. J. Provost, and M. R. Hilleman, "Development of Immunity against Hepatitis A Virus by Subclinical Infection," *Proceedings of the Society for Experimental Biology and Medicine* 153 (1976): 205–8; P. J. Provost, V. M. Villarejos, and M. R. Hilleman, "Suitability of the Rufiventer Marmoset as a Host Animal for Human Hepatitis A Virus," *Proceedings of the Society for Experimental Biology and Medicine* 155 (1977): 283–86; P. J. Provost, V. M. Villarejos, and M. R. Hilleman, "Tests in Rufiventer and Other Marmosets of Susceptibility to Human Hepatitis A Virus," *Primates in Medicine* 10 (1978): 288–94; P. J. Provost and M. R. Hilleman, "An Inactivated Hepatitis A Vaccine Virus Prepared from Infected Marmoset Liver," *Proceedings of the Society for Experimental Biology and Medicine* 159 (1978): 201–3; P. J. Provost and M. R. Hilleman, "Propagation of Human Hepatitis A Virus in Cell Culture in vitro," *Proceedings of the Society for Experimental Biology and Medicine* 160 (1979): 213–21; P. J. Provost, P. A. Giesa, W. J. McAleer, and M. R. Hilleman, "Isolation of Hepatitis A Virus in vitro in Cell Culture Directly from Human Specimens," *Proceedings of the Society for Experimental Biology and Medicine* 167 (1981): 201–6; P. J. Provost, F. S. Banker, P. A. Giesa, W. J. McAleer, E. B. Buynak, and M. R. Hilleman, "Progress Toward a Live, Attenuated Human Hepatitis A Vaccine," *Proceedings of the Society for Experimental Biology and Medicine* 170 (1982): 8–14; P. J. Provost, P. A. Conti, P. A. Giesa, F. S. Banker, E. B. Buynak, W. J. McAleer, and M. R. Hilleman, "Studies in Chimps of Live, Attenuated Hepatitis A Vaccine Candidates," *Proceedings of the Society for Experimental Biology and Medicine* 172 (1983): 357–63; W. M. Hurmi, W. J. Miller, W. J. McAleer, P. J. Provost, and M. R. Hilleman, "Viral Enhancement and Interference Induced in Cell Culture by Hepatitis A Virus: Application to Quantitative Assays for Hepatitis A Virus," *Proceedings of the Society for Experimental Biology and Medicine* 175 (1984): 84–87.

**107** Kiryas Joel: M. Hill, "Hasidic Enclave Has Growing Pains in Suburbia," Associated Press, September 11, 2004.

**108** Werzberger hepatitis A vaccine study: A. Werzberger, B. Mensch, B. Kuter, L. Brown, J. Lewis, R. Sitrin, W. Miller, D. Shouval, B. Wiens, G. Calandra, J. Ryan, P. Provost, and D. Nalin, "A Controlled Trial of a Formalin-Inactivated Hepatitis A Vaccine in Healthy Children," *New England Journal of Medicine* 327 (1992): 453–57.

111 Leonard Hayflick persecution: P. M. Boffey, "The Fall and Rise of Leonard Hayflick," *New York Times*, January 19, 1982; L. Hayflick, "WI–38: From Purloined Cells to National Policy," *Current Contents*, January 15, 1990; N. Wade, "Hayflick's Tragedy: The Rise and Fall of a Human Cell Line," *Science* 192 (1976): 125–27; C. Holden, "Hayflick Case Settled," *Science* 215 (1982): 271; B. L. Strehler, "Hayflick-NIH Settlement," *Science* 215 (1982): 240–42.

## Blood

Joan Staub and Bert Peltier were interviewed on May 15, 2006, and January 11, 2005, respectively.

113 AIDS cluster: D. M. Auerbach, W. W. Darrow, H. W. Jaffe, and J. W. Curran, "Cluster of Cases of Acquired Immune Deficiency Syndrome: Patients Linked by Sexual Contact," *American Journal of Medicine* 76 (1984):4 87–92.

114 Gaetan Dugas: R. Shilts, *And the Band Played On* (New York: St. Martin's Press, 1987).

115 Hepatitis B disease: Plotkin and Orenstein, *Vaccines*.

117 Blumberg background: B. S. Blumberg, "The Discovery of the Hepatitis B Virus and the Invention of the Vaccine: A Scientific Memoir," *Journal of Gastroenterology and Hepatology* 17 (2002): S502; Hall of Fame, Inventor Profile, "Vaccine against Viral Hepatitis and Process: Process of Viral Diagnosis and Reagent Vaccine for Hepatitis B," http://www.invent.org.hall_of_fame/17.html; P. Wortsman, "Profile: Baruch Blumberg '51," *P & S Journal* 16, no. 1 (Winter 1996); F. Blank, "76 Revolutionary Minds," phillymag.com, http://www.phillymag.com/Archives/ 2001Nov/smart_2.html; "Baruch S. Blumberg," http://britannica.com/nobel/micro/74_63.html; "Baruch S. Blumberg: Autobiography," http://nobelprize.org/medicine/laureates/1976/Blumberg-autobio.html; "The Hepatitis B Story," http://www.beyond discovery.org.

118 Blumberg hepatitis studies: B. S. Blumberg, "Polymorphisms of the Serum Proteins and the Development of Iso-Precipitins in Transfused Patients," *Bulletin of the New York Academy of Medicine* 40 (1964): 377–86; B. S. Blumberg, J. S. Gerstley, D. A. Hungerford, et al., "A Serum Antigen (Australia Antigen) in Down's Syndrome, Leukemia, and Hepatitis," *Annals of Internal Medicine* 66 (1967): 924–31; B. S. Blumberg, A. I. Sutnick, and W. T. London, "Hepatitis and Leukemia: Their Relation to Australia Antigen," *Bulletin of the New York Academy of Medicine* 44 (1968): 1566–86; B. S. Blumberg, "Australia Antigen and the Biology of Hepatitis B: Nobel Lecture," December 13, 1976.

119 Prince hepatitis study: A. M. Prince, "An Antigen Detected in the Blood During the Incubation Period of Serum Hepatitis," *Proceedings of the National Academy of Sciences* 60 (1968): 814–21.

120 Krugman background: Saul Krugman: Physician, scientist, teacher,

1911–1995, http://library.med.nyu.edu/library/eresources/featuredcollections/krugman/html.

120 Krugman Willowbrook studies: S. Krugman and R. Ward, "Clinical and Experimental Studies of Infectious Hepatitis," *Pediatrics* 22 (1958): 1016–22; S. Krugman, J. P. Giles, and J. Hammond, "Viral Hepatitis, Type B (MS–2 Strain): Studies on Active Immunization," *Journal of the American Medical Association* 217 (1971): 41–45; S. Krugman, J. P. Giles, and J. Hammond, "Hepatitis Virus: Effect of Heat on the Infectivity and Antigenicity of the MS–1 and MS–2 Strains," *Journal of Infectious Diseases* 122 (1970): 432–36; S. Krugman, "The Willowbrook Hepatitis Studies Revisited: Ethical Aspects," *Reviews of Infectious Diseases* 8 (1986): 157–62.

121 Seymour Thaler: Radetsky, *Invaders.*

122 Hilleman blood-derived hepatitis B vaccine studies: M. R. Hilleman, E. B. Buynak, R. R. Roehm, et al., "Purified and Inactivated Human Hepatitis B Vaccine," *American Journal of the Medical Sciences* 270 (1975): 401–4; E. B. Buynak, R. R. Roehm, A. A. Tytell, A. U. Bertland, G. P. Lampson, and M. R. Hilleman, "Development and Chimpanzee Testing of a Vaccine against Human Hepatitis B," *Proceedings of the Society for Experimental Biology and Medicine* 151 (1976): 694–700; E. B. Buynak, R. R. Roehm, A. A. Tytell, A. U. Bertland, G. P. Lampson, and M. R. Hilleman, "Vaccine against Human Hepatitis B," *Journal of the American Medical Association* 235 (1976): 2832–34; E. Tabor, E. Buynak, L. A. Smallwood, P. Snoy, M. Hilleman, and R. Gerety, "Inactivation of Hepatitis B Virus by Three Methods: Treatment with Pepsin, Urea, or Formalin," *Journal of Medical Virology* 11 (1983): 1–9.

125 Alter quote: Radetsky, *Invaders.*

128 Vagelos quote: Symposium in honor of Maurice R. Hilleman, American Philosophical Society, January 26, 2005.

128 Jeryl Hilleman quote: Ibid.

132 Wolf Szmuness: Radetsky, *Invaders.*

132 Szmuness hepatitis B vaccine study: W. Szmuness, C. E. Stevens, E. J. Harley, E. A. Zang, W. R. Oleszko, D. C. William, R. Sadovsky, J. M. Morrison, and A. Kellner, "Hepatitis B Vaccine: Demonstration of Efficacy in a Controlled Clinical Trial in a High-Risk Population in the United States," *New England Journal of Medicine* 303 (1980): 833–41.

133 Cantwell: A. Cantwell Jr., *AIDS and the Doctors of Death* (Los Angeles: Aries Rising Press, 1988).

133 Relationship between hepatitis B vaccine and AIDS: "Current Trends in Hepatitis B Vaccine: Evidence Confirming Lack of AIDS Transmission," *Morbidity and Mortality Weekly Report* 33 (1984): 685–87.

137 Boyer and Cohen: "Biotechnology at 25: The Founders," http://bancroft.berkeley.Edu/Exhibits/Biotech/25.html; "Robert Swanson and Herbert Boyer: Giving Birth to Biotech," *BusinessWeek Online*, http://www.businessweek.com/magazine/content/04_42/b3904017_mz072.htm; "A Historical Timeline: Cracking the Code of Life," http://www.jgi.doe.gov/education/timeline_3.html; "Who Made America?: Herbert Boyer: Biotechnology," http://www.

pbs.org/wgbh/theymadeamerica/whomade/boyer_hi.html; 1973, "Herbert Boyer (1936–) and Stanley Cohen (1936–) Develop Recombinant DNA Technology, Showing That Genetically Engineered DNA Molecules May Be Cloned in Foreign Cells," *Genome News Network*, http://www.genomenewsnetwork.org/resources/timeline/1973_Boyer.php; "Shaping Life in the Lab: The Boom in Genetic Engineering," *Time*, March 9, 1981; "Herbert Boyer (1936–)," http://www.accessexcellence.org/RC/AB/BC/Herbert_ Boyer.html; "The Birth of Biotech," *Technology Review.com*; http://www.technologyreview.com/articles/00/07/trailing0700.asp; "Stanley Cohen and Herbert Boyer," http://www.nobel-prize-winners.com/cohen/cohen.html; "Herbert W. Boyer, PhD," Forbes.com; Inventor of the Week Archive, "Cloning of Genetically Engineered Molecules," http://web.mit.edu/invent/iow/boyercohen.html.

139 Hilleman recombinant hepatitis B vaccine studies: P. Valenzuela, A. Medina, W. J. Rutter, G. Ammerer, and B. D. Hall, "Synthesis and Assembly of Hepatitis B Virus Surface Antigen Particles in Yeast," *Nature* 298 (1982): 347–50; W. J. McAleer, E. B. Buynak, R. Z. Maigetter, D. E. Wampler, W. J. Miller, and M. R. Hilleman, "Human Hepatitis B Vaccine from Recombinant Yeast," *Nature* 307 (1984): 178–80; M. R. Hilleman, R. E. Weibel, and E. M. Scolnick, "Recombinant Yeast Human Hepatitis B Vaccine," *Journal of the Hong Kong Medical Association* 37 (1985): 75–85; M. R. Hilleman, R. E. Weibel, and E. M. Scolnick, "Research on Hepatitis B Vaccine Continues Unabated," *Medical Progress*, August 1985: 49–51; M. R. Hilleman, "Recombinant Yeast Hepatitis B Vaccine," *Development of Biological Standards* 63 (1986): 57–62; M. R. Hilleman and R. Ellis, "Vaccines Made from Recombinant Yeast Cells," *Vaccine* 4 (1986): 75–76; M. R. Hilleman, "Present and Future Control of Human Hepatitis B by Vaccination, in *Modern Biotechnology and Health: Perspectives for the Year 2000* (New York: Academic Press, 1987); M. R. Hilleman, "Present Status of Recombinant Hepatitis B Vaccine," *Acta Paediatrica Scandinavica* 29 (1988): 8B–15B; M. R. Hilleman, "Vaccines in Perspective: Human Hepatitis B Vaccines, the First Subunit Recombinant Viral Vaccines," in *Current Topics in Biomedical Research*, ed. R. Kurth and W. K. Schwerdtfeger (Berlin and Heidelberg: Springer-Verlag, 1992), 145–61; M. R. Hilleman, "Vaccine Perspectives from the Vantage of Hepatitis B," *Vaccine Research* 1 (1992): 1–15; M. R. Hilleman, "Three Decades of Hepatitis Vaccinology in Historic Perspective: A Paradigm for Successful Pursuits," in Plotkin and Fantini, *Vaccinia*; M. R. Hilleman, "Critical Overview and Outlook: Pathogenesis, Prevention, and Treatment of Hepatitis and Hepatocellular Carcinoma caused by Hepatitis B virus," *Vaccine* 21 (2003): 4626–49.

140 Impact of hepatitis B vaccine: Plotkin and Orenstein, *Vaccines*; "Hepatitis B Vaccination Coverage among Adults: United States, 2004," *Morbidity and Mortality Weekly Report* 55 (2006): 509–11.

140 Starzl quote: Symposium in honor of Maurice R. Hilleman, American Philosophical Society, January 26, 2005.

### Animalcules

Robert Austrian was interviewed on February 18, 2005, and September 14, 2006. Much of the information about pneumococcus and pneumococcal vaccine can be found in Austrian, *Pneumococcus*.

**141** Discovery of gold in South Africa: "Three Georges Strike Paydirt," http://www.joburg.org.za/facts/georges.stm; "Joburg's Hidden History," http://www.goldreefcity.co.za/theme_park/joburgs_hidden_history.asp; "Johannesburg: History," http://www.southafrica-travel.net/north/a1johb01.htm.

**142** Krüger: "Stephanus Johannes Paulus Kruger," http://en.wikipedia.org/wiki/President_Kruger.

**142** Pneumococcal pneumonia in gold miners: Austrian, *Pneumococcus*.

**144** Robert Koch: D. S. Burke, "Of Postulates and Peccadilloes: Robert Koch and Vaccine (Tuberculin) Therapy for Tuberculosis," *Vaccine* 11 (1993): 795–804; "Robert Koch," http://www.historylearningsite.co.uk/robert_koch.htm; "Robert Koch: Biography," Nobelprize.org, http://nobelprize.org/medicine/laureates/1905/koch-bio.html.

**145** Roux, Yersin, Behring, and Ramon: Plotkin and Orenstein, *Vaccines*; Plotkin and Fantini, *Vaccinia*.

**146** Thalidomide: R. Brynner and T. Stephens, *Dark Remedy: The Impact of Thalidomide and Its Revival as a Vital Medicine* (New York: Perseus Publishing, 2001).

**146** Food, Drug and Cosmetic Act amendments: "The Story of the Laws behind the Labels: Part II. 1938—The Federal Food, Drug, and Cosmetic Act, Part III. 1962 Drug Amendments," *FDA Consumer*, June 1981.

**147** Almroth Wright: M. Dunhill, *The Plato of Praed Street: The Life and Times of Almroth Wright* (London: Royal Society of Medicine Press, 2002); "The Life and Times of Almroth Wright," *Biomedical Scientist*, March 2002.

**149** Gerhard Domagk: "Gerhard Domagk," NobelPrize.org, http://nobelprize.org.medicine/laureates/1939/domagk-bio.html.

**149** Perrin Long: J. F. Worthington, "The Guys and Us," *Hopkins Medicine Magazine*, Spring/Summer 2005.

**151** Colin MacLeod pneumococcal vaccine studies: C. M. MacLeod, R. G. Hodges, M. Heidelberger, and W. G. Bernhard, "Prevention of Pneumococcal Pneumonia by Immunization with Specific Capsular Polysaccharides," *Journal of Experimental Medicine* 82 (1945): 445–65.

### An Uncertain Future

Adel Mahmoud and Art Caplan were interviewed on May 26, 2006, and March 10, 2005, respectively. An excellent summary of Andrew Wakefield, the MMR vaccine, and autism can be found in Fitzpatrick, *MMR and Autism*.

**156** HPV vaccine: R. Steinbrook, "The Potential of Human Papillomavirus Vaccines," *New England Journal of Medicine* 354 (2006): 1109–12; M. Schiffman and P. E. Castle, "The Promise of Global Cervical-Cancer Prevention," *New England Journal of Medicine* 353(2005): 2101–4; C. P. Crum, "The Beginning of the End for Cervical Cancer?" *New England Journal of*

*Medicine* 34 (2002):1703–5; L. A. Koutsky, K. A. Ault, C. M. Wheeler, et al., "A Controlled Trial of a Human Papillomavirus Type 16 Vaccine," *New England Journal of Medicine* 347 (2002): 1645–51.

**158** "Fry on the tarmac": Interview with senior executive at Merck, 2004.

**158** Bill and Melinda Gates Foundation: S. Okie, "Global Health: The Gates-Buffett Effect," *New England Journal of Medicine* 355 (2006): 1084–88.

**159** Hilleman MMR studies: E. B. Buynak, R. E. Weibel, J. E. Whitman, J. Stokes Jr., and M. R. Hilleman, "Combined Live Measles, Mumps, and Rubella Vaccines," *Journal of the American Medical Association* 207 (1969): 2259–62; R. E. Weibel, J. Stokes Jr., V. M. Villarejos, J. A. Arguedas, E. B. Buynak, and M. R. Hilleman, "Combined Live Rubella-Mumps Virus Vaccine: Findings in Clinical-Laboratory Studies," *Journal of the American Medical Association* 216 (1971): 983–86; J. Stokes Jr., R. E. Weibel, V. M. Villarejos, J. A. Arguedas, E. B. Buynak, and M. R. Hilleman, "Trivalent Combined Measles-Mumps-Rubella Vaccine: Findings in Clinical-Laboratory Studies," *Journal of the American Medical Association* 218 (1971): 57–61; V. M. Villarejos, J. A. Arguedas, E. B. Buynak, R. E. Weibel, J. Stokes Jr., and M. R. Hilleman, "Combined Live Measles-Rubella Vaccine: Findings in Clinical-Laboratory Studies," *Journal of Pediatrics* 79 (1971): 599–604; R. E. Weibel, E. B. Buynak, J. Stokes Jr., and M. R. Hilleman, "Measurement of Immunity Following Live Mumps (5 Years), Measles (3 Years), and Rubella (2 ½ Years) Virus Vaccines," *Pediatrics* 49 (1972): 334–41; J. M. Borgoño, R. Greiber, G. Solari, F. Concha, B. Carrillo, and M. R. Hilleman, "A Field Trial of Combined Measles-Mumps-Rubella Vaccine: Satisfactory Immunization with 188 Children in Chile," *Clinical Pediatrics* 12 (1973): 170–72; R. E. Weibel, V. M. Villarejos, G. Hernández, J. Stokes Jr., E. B. Buynak, and M. R. Hilleman, "Combined Live Measles-Mumps Virus Vaccine," *Archives of Disease in Childhood* 48 (1973): 532–36; R. E. Weibel, E. B. Buynak, J. Stokes Jr., and M. R. Hilleman, "Persistence of Immunity Following Monovalent and Combined Live Measles, Mumps, and Rubella Virus Vaccines," *Pediatrics* 51 (1973): 467–75; R. E. Weibel, E. B. Buynak, A. A. McLean, and M. R. Hilleman, "Long-Term Follow-Up for Immunity after Monovalent or Combined Measles, Mumps, and Rubella Virus Vaccines," *Pediatrics* 56 (1975): 380–87; R. E. Weibel, E. B. Buynak, A. A. McLean, and M. R. Hilleman, "Persistence of Antibody after Administration of Monovalent and Combined Live Attenuated Measles, Mumps, and Rubella Virus Vaccines," *Pediatrics* 61 (1978): 5–11; R. E. Weibel, E. B. Buynak, A. A. McLean, and M. R. Hilleman, "Follow-Up Surveillance for Antibody in Human Subjects Following Live Attenuated Measles, Mumps, and Rubella Virus Vaccines," *Proceedings of the Society for Experimental Biology and Medicine* 162 (1979): 328–32; W. J. McAleer, H. Z. Markus, A. A. McLean, E. B. Buynak, and M. R. Hilleman, "Stability on Storage at Various Temperatures of Live Measles, Mumps, and Rubella Virus Vaccines in New Stabilizer," *Journal of Biological Standardization* 8 (1980): 281–87; R. E. Weibel, E. B. Buynak, A. A. McLean, R. R. Roehm, and M. R. Hilleman, "Persistence of Antibody in

Human Subjects for 7 to 10 Years Following Administration of Combined Live Attenuated Measles, Mumps, and Rubella Virus Vaccines," *Proceedings of the Society for Experimental Biology and Medicine* 165 (1980): 260–63; R. E. Weibel, A. J. Carlson, V. M. Villarejos, E. B. Buynak, A. A. McLean, and M. R. Hilleman, "Clinical and Laboratory Studies of Combined Live Measles, Mumps, and Rubella Vaccines Using the RA27/3 Rubella Virus," *Proceedings of the Society for Experimental Biology and Medicine* 165 (1980): 323–26.

**159** Wakefield study in *Lancet*: A. J. Wakefield, S. H. Murch, A. Anthony, et al., "Ileal-Lymphoid-Nodular Hyperplasia, Non-Specific Colitis, and Pervasive Developmental Disorder in Children," *Lancet* 351 (1998): 637–41.

**159** Wakefield description: Fitzpatrick, *MMR and Autism.*

**160** Shermer quote: Shermer, *Weird Things.*

**161** Four-month-old child with seizures prior to vaccine: Interview with nurse practitioner at Kids First-Haverford pediatric practice, 2004.

**163** Deaths from measles following Wakefield report: Public Health Laboratory Service, "Measles Outbreak in London," *Communicable Disease Report CDR Weekly* 12 (2002): 1; T. Peterkin, "Alert over 60 Percent Rise in Measles," *London Daily Telegraph*, May 12, 2003; B. Lavery, "As Vaccination Rates Decline in Ireland, Cases of Measles Soar," *New York Times*, February 8, 2003; *Fragile Immunity*, video produced by PATH, narrated by Ian Holm, 2004.

**163** Burton hearings: *Autism: Present Challenges, Future Needs: Why the Increased Rates?* Hearing before the Committee on Government Reform, House of Representatives, 106th Congress, 2d Session (Washington, D.C.: U.S. Government Printing Office, 2001).

**167** MMR-autism studies: R. T. Chen and F. De Stefano, "Vaccine Adverse Events: Causal of Coincidental?" *Lancet* 351 (1968): 611–12; B. Taylor, E. Miller, C. P. Farrington, et al., "Autism and Measles, Mumps, and Rubella Vaccine: No Epidemiological Evidence for a Causal Association," *Lancet* 353 (1999): 2026–29; C. P. Farrington, E. Miller, and B. Taylor, "MMR and Autism: Further Evidence against a Causal Association," *Vaccine* 19 (2001): 3632–35; R. L. Davis, P. Kramarz, K. Bohlke, et al., "Measles-Mumps-Rubella and Other Measles-Containing Vaccines Do Not Increase the Risk for Inflammatory Bowel Disease: A Case-Control Study from the Vaccine Safety Datalink Project," *Archives of Pediatric and Adolescent Medicine* 155 (2001): 354–59; J. A. Kaye, M. Melero-Montes, and H. Jick, "Mumps, Measles, and Rubella Vaccine and the Incidence of Autism Recorded by General Practitioners: A Time Trend Analysis," *British Medical Journal* 322 (2001): 460–63; L. Dales, S. J. Hammer, and N. J. Smith, "Time Trends in Autism and in MMR Immunization Coverage in California," *Journal of the American Medical Association* 285 (2001): 1183–85; E. Fombonne and S. Chakrabarti, "No Evidence for a New Variant of Measles-Mumps-Rubella-Induced Autism," *Pediatrics* 108 (2001): E58; K. Stratton, A. Gable, and P. M. M. Shetty, ed. *Measles-Mumps-Rubella Vaccine and Autism, in Immunization Safety Review*, Institute of Medicine, Immunization Safety Review Committee

(Washington, D.C.: National Academy Press, 2001); B. Taylor, E. Miller, R. Lingam, et al., "Measles, Mumps, and Rubella Vaccination and Bowel Problems or Developmental Regression in Children with Autism: Population Study," *British Medical Journal* 324 (2002): 393–96; K. Wilson, E. Mills, C. Ross, et al., "Association of Autistic Spectrum Disorder and the Measles, Mumps, and Rubella Vaccine: A Systematic Review of Current Epidemiological Evidence," *Archives of Pediatric and Adolescent Medicine* 157 (2003): 628–34; E. Miller, "Measles-Mumps-Rubella Vaccine and the Development of Autism," *Seminars in Pediatric Infectious Diseases* 14 (2003): 199–206; K. M. Madsen and M. Vestergaard, "MMR Vaccination and Autism: What Is the Evidence for a Causal Association?" *Drug Safety* 27 (2004): 831–40; F. DeStefano and W. W. Thompson, "MMR Vaccine and Autism: An Update of the Scientific Evidence," *Expert Reviews in Vaccines* 3 (2004): 19–22; F. DeStefano, T. K. Bhasin, W. W. Thompson, et al., "Age at First Measles-Mumps-Rubella Vaccination in Children with Autism and School-Matched Control Subjects: A Population-Based Study in Metropolitan Atlanta," *Pediatrics* 113 (2004): 259–66; H. Honda, Y. Shimizu, and M. Rutter, "No Effect of MMR Withdrawal on the Incidence of Autism: A Total Population Study," *Journal of Child Psychology and Psychiatric Allied Disciplines* 46 (2005): 572–79.

**167** Brian Deer and the press: Fitzpatrick, *MMR and Autism*; B. Deer, "MMR Scare Doctor Faces List of Charges," www.timesonline.co.uk/article/0.,2087–1774388,00.html; K. Birmingham and M. Cimons, "Reactions to MMR Immunization Scare," *Nature Medicine Vaccine Supplement* 4 (1998): 478–79; G. Crowley and G. Brownwell, "Parents Wonder: Is It Safe to Vaccinate?" *Newsweek*, July 31, 2000; K. Seroussi, "We Cured Our Son's Autism," *Parents*, February 2000; D. Brown, "Autism's New Face," *Washington Post*, March 26, 2000; "Rash Worries," *The Economist*, April 11, 1998; A. Manning, "Vaccine-Autism Link Feared," *USA Today*, August 19, 1999; P. Anderson, "Another Media Scare about MMR Vaccine Hits Britain," *British Medical Journal*, June 12, 1999; B. Vastag, "Congressional Autism Hearings Continue: No Evidence MMR Vaccine Causes Disorder," *Journal of the American Medical Association* 285 (2001): 2567–69; "Does the MMR Vaccine Cause Autism?" *Mothering*, September/October 1998; S. Ramsey, "UK Starts Campaign to Reassure Parents about MMR-Vaccine Safety," *Lancet* 357 (2001): 290; N. Bragg, M. Ramsay, J. White, and Z. Bozoky, "Media Dents Confidence in MMR Vaccine," *British Medical Journal* 316 (1998): 561; J. Fischman, "Vaccine Worries Get Shot Down but Parents Still Fret," *U.S. News and World Report*, March 19, 2001.

**170** Mercury and autism: Excellent summary of current studies in the Institute of Medicine's *Immunization Safety Review: Vaccines and Autism* (Washington, D.C.: National Academy Press, May 17, 2004), and *Immunization Safety Review: Thimerosal-Containing Vaccines and Neurodevelopmental Disorders* (Washington, D.C.: National Academy Press, October 1, 2001).

**172** AAP-PHS statement on thimerosal: American Academy of Pediatrics. Committee on Infectious Diseases and Committee on Environmental Health, "Thimerosal in Vaccines: An Interim Report to Clinicians," *Pediatrics* 104 (1999): 570–74.

173 Kennedy *Rolling Stone* article: R. F. Kennedy Jr., "Deadly Immunity," *Rolling Stone*, June 20, 2005.

173 Schwarzenegger bans thimerosal: J. S. Lyon, "Dearth of Vaccines for Infants and Experts Urge Use for First Time," *San Jose Mercury News*, November 1, 2004.

174 Evidence of Harm: D. Kirby, *Evidence of Harm: Mercury in Vaccines and the Autism Epidemic: A Medical Controversy* (New York: St. Martin's Press, 2005).

174 Chelation death: K. Kane and V. Linn, "Boy Dies During Autism Treatment," *Pittsburgh Post-Gazette*, August 25, 2005.

175 Hilleman memo: M. Levin, "'91 Memo Warned of Mercury in Shots," *Los Angeles Times*, February 8, 2005.

176 Thimerosal-autism studies: A. Hviid, M. Stellfeld, J. Wohlfahrt, and M. Melbye, "Association between Thimerosal-Containing Vaccine and Autism," *Journal of the American Medical Association* 290 (2003): 1763–66; T. Verstraeten, R. L. Davis, F. DeStefano, et al., "Safety of Thimerosal-Containing Vaccines: A Two-Phased Study of Computerized Health Maintenance Organization Databases," *Pediatrics* 112 (2003): 1039–48; J. Heron, J. Golding, and ALSPAC Study Team, "Thimerosal Exposure in Infants and Developmental Disorders: A Prospective Cohort Study in the United Kingdom Does Not Show a Causal Association," *Pediatrics* 114 (2004): 577–83; N. Andrews, E. Miller, A. Grant, et al., "Thimerosal Exposure in Infants and Developmental Disorders: A Retrospective Cohort Study in the United Kingdom Does Not Show a Causal Association," *Pediatrics*, 114 (2004):584–91; S. Parker, B. Schwartz, J. Todd, and L. K. Pickering, "Thimerosal-Containing Vaccines and Autistic Spectrum Disorder: A Critical Review of Published Original Data," *Pediatrics* 114 (2004): 793–804; E. Fombonne, R. Zakarian, A. Bennett, et al., "Pervasive Developmental Disorders in Montreal, Quebec, Canada: Prevalence and Links with Immunization," *Pediatrics* 118 (2006): 139–50.

177 Choosing not to afford vaccines: J. Cohen, "U.S. Vaccine Supply Falls Seriously Short," *Science* 295 (2002): 1998–2001; National Vaccine Advisory Committee, "Strengthening the Supply of Routinely Recommended Vaccines in the United States: Recommendations of the National Vaccine Advisory Committee," *Journal of the American Medical Association* 290 (2003): 3122–28; Committee on the Evaluation of Vaccine Purchase Financing in the United States, Institute of Medicine, *Financing Vaccines in the 21st Century: Assuring Access and Availability* (Washington, D.C.: National Academy Press, 2004); S. Stokley, K. M. Shaw, L. Barker, J. M. Santoli, and A. Shefer, "Impact of State Vaccine Financing Policy on Uptake of Heptavalent Pneumococcal Conjugate Vaccine," *American Journal of Public Health* 96 (2006): 1308–13.

**178** *John Q*: New Line Cinema, 2002

**179** HPV vaccine controversy: J. Guyton, The coming storm over a cancer vaccine, *Fortune*, October 31, 2005.

**180** Vaccine profitability: S. Garber, *Product Liability and the Economics of Pharmaceuticals and Medical Devices* (Santa Monica: RAND, Institute for Civil Justice, 1993); R. Manning: "Economic Impact of Product Liability in U.S. Prescription Drug Markets," *International Business Lawyer* March 2001, and "Changing Rules in Tort Law and the Market for Childhood Vaccines," *Journal of Law and Economics* 37 (1994): 247–75; Institute of Medicine, *Financing Vaccines in the 21st Century: Assuring Access and Availability* (Washington, D.C.: National Academy Press, 2004); J. M. Wood, "Litigation Could Make Vaccines Extinct," *The Scientist*, January 19, 2004; T. Ginsberg, "Making Vaccines Worth It," *Philadelphia Inquirer*, September 24, 2006.

**182** Measles outbreak in John Hancock Tower: R. Know, "Measles Outbreak Shows Even Vaccinated at Risk," *National Public Radio*, June 21, 2006.

**182** Mumps epidemic in Midwest: Centers for Disease Control and Prevention, "Mumps Epidemic: Iowa, 2006," *Morbidity and Mortality Weekly Report* 55 (2006): 366–68.

**183** Polio outbreak in The Netherlands: P. M. Oostvogel, J. K. van Wijngaarden, H. G. van der Avoort, et al., "Poliomyelitis Outbreak in an Unvaccinated Community in The Netherlands: 1992–93," *Lancet* 344 (1994): 665–70; Centers for Disease Control and Prevention, "Follow-Up on Poliomyelitis—United States, Canada, Netherlands: 1979," *Morbidity and Mortality Weekly Report* 46 (1997): 1195–99; H. C. Rumke, P. M. Oostvogel, G. Van Steenis, and A. M. Van Loon, "Poliomyelitis in The Netherlands: A Review of Population Immunity and Exposure between the Epidemics in 1978 and 1992," *Epidemiology and Infection* 115 (1995): 289–98; H. Bijkerk, "Poliomyelitis Epidemic in the Netherlands: 1978," *Developments in Biological Standardization* 43 (1979): 195–206.

**183** Diphtheria outbreak in the former Soviet Union: Centers for Disease Control and Prevention, "Update: Diphtheria Epidemic—New Independent States of the Former Soviet Union, January 1995–March 1996," *Morbidity and Mortality Weekly Report* 45 (1996): 693–97.

### Unrecognized Genius

**186** Strauss quote: Walter Strauss, personal communication, September 20, 2006.

**188** Leningrad and Urabe strains of mumps vaccine: Plotkin and Orenstein, *Vaccines*.

**188** Hilleman obituary: L. Altman, "Maurice Hilleman, Master in Creating Vaccines, Dies at 85," *New York Times*, April 12, 2005.

**189** Fauci quote: "The Vaccine Hunter," BBC Radio 4, producer Pauline Moffatt, June 21, 2006.

**190** Strauss quote: Walter Strauss, personal communication, September 20, 2006.

190 Röntgen: B. Goldsmith, *Obsessive Genius: The Inner World of Marie Curie* (New York: W.W. Norton, 2005).

190 Sam Katz correction of *Rolling Stone* article: Salon.com, July 21, 2005.

191 Salk and patent: J. Smith, *Patenting the Sun*.

191 *Twister*: Warner Brothers, 1996.

191 Robert Gallo quote: "The Vaccine Hunter," BBC Radio 4, producer Pauline Moffatt, June 21, 2006.

192 Lorraine Hilleman quote: Interview, September 14, 2006.

192 Jeryl Hilleman regarding photograph: Interview, March 11, 2005.

192 Szmuness paper: W. Szmuness, C. D. Stevens, E. J. Harley, E. A. Zang, W. R. Oleszko, D. C. William, R. Sadovsky, J. M. Morrison, and A. Kellner, "Hepatitis B Vaccine: Demonstration of Efficacy in a Controlled Clinical Trial in a High-Risk Population in the United States," *New England Journal of Medicine* 303 (1980): 833–41.

193 Measles outbreak, 1989–1991: Centers for Disease Control, "Public-Sector Vaccination Efforts in Response to the Resurgence of Measles among Preschool-Aged Children: United States, 1989–1991," *Morbidity and Mortality Weekly Report* 41 (1992): 522–25.

194 Alfred Nobel and the Nobel Prize: "Alfred Nobel: The Man," http://www.britannica.com/nobel/micro/427_33.html; "Excerpt from the Will of Alfred Nobel," Nobelprize.org, http://nobelprize.org.nobel/alfred-nobel/biographical/will/ index.html; "Alfred Nobel," http://en.wikipedia.irg/wiki/Alfred_Nobel.

196 Salk and Eisenhower: J. Smith, *Patenting the Sun*.

196 Salk obituary: R. Dulbecco, "Jonas Salk," *Nature* 376 (1995): 216.

197 Undeserving Nobel Prizes: L. K. Altman, "Alfred Nobel and the Prize That Almost Didn't Happen," *New York Times*, September 26, 2006.

199 Fauci quote: "The Vaccine Hunter," BBC Radio 4, producer Pauline Moffatt, June 21, 2006.

## Epilogue

Roy Vagelos, Anthony Fauci, Lorraine Hilleman, Jeryl Hilleman, and Kirsten Hilleman all spoke at the symposium in Hilleman's honor at the American Philosophical Society, January 26, 2005.

201 American Philosophical Society: American Philosophical Society, http://www.amphilsoc.org; American Philosophical Society Library and Museum, http://www.ushistory.org/tour/tour_philo.htm.

205 Heat-shock proteins: E. Gilboa, "The Promise of Cancer Vaccines," *Nature Reviews* 4 (2004): 401–11; R. Suto and P. K. Srivastava, "A Mechanism for the Specific Immunogenicity of Heat-Shock Protein–Chaperoned Peptides," *Science* 269 (1995): 1585–88; A. Hoos and D. L. Levey, "Vaccination with Heat-Shock Protein-Peptide Complexes: From Basic Science to Clinical Applications," *Expert Reviews of Vaccines* 2 (2003): 369–79; P. K. Srivastava and M. R. Das, "Serologically Unique Surface Antigen of a Rat Hepatoma

Is Also Its Tumor-Associated Transplant Antigen," *International Journal of Cancer* 33 (1984): 417–22.

206 "Pass around a cup" quote: H. Collins, "The Man Who Changed Your Life," *Philadelphia Inquirer*, August 29, 1999.

206 Faulkner quote: "William Faulkner, Nobel Prize Acceptance Speech, Stockholm, Sweden, December 10, 1950," http://www.rjgeib.com/thoughts/faulkner/faulkner.html.

# Selected Bibliography

Angell, Marsha. *Science on Trial: The Clash of Medical Evidence and the Law in the Breast Implant Case*. New York: W. W. Norton and Company, 1996.

Austrian, Robert. *Life with the Pneumococcus: Notes from the Bedside, Laboratory, and Library*. Philadelphia: University of Pennsylvania Press, 1985.

Barry, John. *The Great Influenza: The Epic Story of the Deadliest Plague in History*. New York: Viking, 2004.

Blumberg, Baruch. *Hepatitis B: The Hunt for a Killer Virus*. Princeton: Princeton University Press, 2002.

Bookchin, Debbie, and Jim Schumacher. *The Virus and the Vaccine: The True Story of a Cancer-Causing Monkey Virus, Contaminated Polio Vaccine, and the Millions of Americans Exposed*. New York: St. Martin's Press, 2004.

Carter, Richard. *Breakthrough: The Saga of Jonas Salk*. New York: Trident Press, 1966.

Collins, Robert. *Ernest William Goodpasture: Scientist, Scholar, Gentleman*. Franklin, TN: Hillsboro Press, 2002.

Debré, Patrice. *Louis Pasteur*. Baltimore and London: Johns Hopkins University Press, 1998.

De Kruif, Paul. *Microbe Hunters*. New York: Harcourt, Brace, and Company, 1926.

Etheridge, Elizabeth. *Sentinel for Health: A History of the Centers for Disease Control*. Berkeley: University of California Press, 1992.

Fitzpatrick, Michael. *MMR and Autism: What Parents Need to Know*. London and New York: Routledge, 2004.

Galambos, Louis, and Jane Eliot Sewell. *Networks of Innovation: Vaccine Development at Merck, Sharpe & Dohme, and Mulford, 1895–1995*. Cambridge: Cambridge University Press, 1995.

Geison, Gerald. *The Private Science of Louis Pasteur*. Princeton: Princeton University Press, 1995.

Hall, Stephen. *Merchants of Immortality: Chasing the Dream of Human Life Extension*. Boston: Houghton Mifflin, 2003.

Hayflick, Leonard. *How and Why We Age*. New York: Ballantine Books, 1994.

Hilts, Philip. *Protecting America's Health: The FDA, Business, and One Hundred Years of Regulation*. New York: Alfred A. Knopf, 2003.

Hilts, Philip. *Rx for Survival: Why We Must Rise to the Global Health Challenge*. New York: Penguin Press, 2005.

Holton, Gerald, ed. *The Twentieth Century Sciences: Studies in the Biography of Ideas*. New York: W. W. Norton and Company, 1972.

Huber, Peter. *Galileo's Revenge: Junk Science in the Courtroom*. New York: Basic Books, 1991.

——. *Liability: The Legal Revolution and Its Consequences*. New York: Basic Books, 1988.

Kolata, Gina. *Flu: The Story of the Great Influenza Pandemic of 1918 and the Search for the Virus That Caused It*. New York: Touchstone, 1999.

Koprowski, Hilary, and Michael B.A. Oldstone, eds. *Microbe Hunters: Then and Now*. Bloomington, IL: Medi-Ed Press, 1996.

Lax, Eric. *The Mold in Dr. Florey's Coat: The Story of the Penicillin Miracle*. New York: Henry Holt and Company, 2004.

Leuchtenburg, William. *A Troubled Feast: American Society since 1945*. Boston: Little, Brown and Company, 1973.

Marks, Harry. *The Progress of Experiment: Science and Therapeutic Reform in the United States, 1900–1990*. Cambridge: Cambridge University Press, 1997.

McNeill, William. *Plagues and Peoples*. New York: Anchor Books, 1976.

Offit, Paul. *The Cutter Incident: How America's First Polio Vaccine Led to the Growing Vaccine Crisis*. New Haven and London: Yale University Press, 2005

Oshinsky, David. *Polio: An American Story*. Oxford and New York: Oxford University Press, 2005

Park, Robert. *Voodoo Science: The Road from Foolishness to Fraud*. Oxford: Oxford University Press, 2000.

Plotkin, Stanley A., and Bernardino Fantini, eds. *Vaccinia, Vaccination, Vaccinology: Jenner, Pasteur and Their Successors*. Paris: Elsevier, 1996.

Plotkin, Stanley A., and Walter A. Orenstein, eds. *Vaccines*, 4th ed. Philadelphia: Saunders, 2004.

Radetsky, Peter. *The Invisible Invaders: The Story of the Emerging Age of Viruses*. Boston: Little, Brown and Company, 1991.

Rothman, David, and Sheila Rothman. *The Willowbrook Wars: Bringing the Mentally Disabled into the Community*. New Brunswick, NJ, and London: Transaction, 2005.

Schreibman, Laura. *The Science and Fiction of Autism*. Cambridge, MA, and London: Harvard University Press, 2005.

Shermer, Michael. *Why People Believe Weird Things: Pseudo-Science, Superstition, and Bogus Notions of Our Time*. New York: MJF Books, 1997.

Shorter, Edward. *The Health Century*. New York: Doubleday, 1987.

Smith, Jane. *Patenting the Sun: Polio and the Salk Vaccine*. New York: William Morrow and Company, 1990.

Smith, Page, and Charles Daniel. *The Chicken Book*. Athens: University of Georgia Press, 2000.

Tucker, Jonathan. *Scourge: The Once and Future Threat of Smallpox*. New York: Atlantic Monthly Press, 2001.

van Iterson, G., L.E. Den Dooren De Jong, and A. J. Kluyver. *Martinus Willem Beijerinck: His Life and His Work*. Madison: Science Tech, 1940.

Weller, Thomas. *Growing Pathogens in Tissue Cultures: Fifty Years in Academic Tropical Medicine, Pediatrics, and Virology*. Canton, MA: Scientific History Publications, 2004.

Williams, Greer. *Virus Hunters*. New York: Alfred A. Knopf, 1960.

# ACKNOWLEDGMENTS

I wish to thank Thomas J. Kelleher, senior editor, Smithsonian Books, for his superb editing, humor, and patience; Andrew Zack, for his unfailing belief in and support of this project; Bojana Ristich, for her lessons on style, form, and logic; Nina Long, for her guidance through the Wistar Institute's archival material; David Rose, for his knowledge of the March of Dimes Birth Defects Foundation archives; and Alan Cohen, Brian Fisher, Peggy Flynn, Frank Hoke, Jason Kim, Kenyetta McDonald, Peggy McGratty, Donald Mitchell, Bonnie Offit, Jason Schwartz, Michael Smith, Kirsten Thistle, Amy Wilen, Allison Wahl, and Theo Zaoutis for their careful reading of the manuscript and their helpful suggestions and criticisms.

I would also like to thank Arthur Allen, Robert Austrian, Art Caplan, Mark Feinberg, Penny Heaton, Leonard Hayflick, Jeryl Hilleman, Kirsten Hilleman, Lorraine Hilleman, Samuel Katz, Barbara Kuter, Margaret Liu, Adel Mahmoud, Charlotte Moser, Walter Orenstein, Bert Peltier, Georges Peter, Amy Pisani, Susan Plotkin, Stanley Plotkin, Phil Provost, Adam Ratner, Lance Rodewald, William Schaffner, Anne Schuchat, Keerti Shah, Joan Staub, Walter Strauss, Roy Vagelos, Robert Weibel, David Weiner, Jeffrey Weiser, and Deborah Wexler for their recollections of Maurice Hilleman or their expertise on the science or history of vaccines.

# INDEX

## ABOUT THE AUTHOR

Paul A. Offit, M.D., is the director of the Vaccine Education Center at the Children's Hospital of Philadelphia, as well as the Maurice R. Hilleman Professor of Vaccinology and Professor of Pediatrics at the University of Pennsylvania School of Medicine. A national expert on vaccines and cocreator of the rotavirus vaccine, Dr. Offit is a recipient of many awards, including a Research Career Development Award from the National Institutes of Health. Dr. Offit was elected to the Institute of Medicine of the National Academies and is a member of the FDA's Vaccine Advisory Committee.